Macromolecular Symposia

Symposium Editors: D. Acierno, L. Di Maio

180

WILEY-VCH

pp. 1–266

March 2002

Macromolecular Symposia publishes lectures given at international symposia and is issued irregularly, with normally 14 volumes published per year. For each symposium volume, an Editor is appointed. The articles are peer-reviewed. The journal is produced by photo-offset lithography directly from the authors' typescripts.
Further information for authors can be obtained from:
Editorial office "Macromolecular Symposia"
Wiley-VCH
P. O. Box 10 11 61, 69451 Weinheim,
Germany
or for parcel and courier services: Pappelallee 3, 69469 Weinheim,
Germany
Tel. +49 (0) 62 01/6 06-2 38 or -5 81; Fax +49 (0) 62 01/6 06-3 09 or 5 10; macromol@wiley-vch.de
http://www.wiley-vch.de/home/macrosymp
Suggestions or proposals for conferences or symposia to be covered in this series should also be sent to the Editorial office at the address above.

Macromolecular Symposia:
Annual subscription rates 2002 (print only or online only)*
Germany, Austria € 1198; Switzerland SFr 1968; other Europe € 1198; outside Europe US $ 1448.
Macromolecular Package, including Macromolecular Chemistry & Physics (18 issues), Macromolecular Bioscience (9 issues), Macromolecular Rapid Communications (18 issues), Macromolecular Theory & Simulations (9 issues) is also available. Details on request.
* For a 5 % premium in addition to **Print Only** or **Online Only**, Institutions can also choose both print and online access.
Packages including Macromolecular Symposia and Macromolecular Materials & Engineering are also available. Details on request.
Single issues and back copies are available. Please, inquire for prices.

Orders may be placed through your bookseller or directly at the publishers: WILEY-VCH Verlag GmbH, P. O. Box 10 11 61, 69451 Weinheim, Germany, Tel.: (0 62 01) 6 06–0, Telefax (0 62 01) 60 61 17, Telex 46 55 16 vchwh d. E-mail: subservice@wiley-vch.de

Macromolecular Symposia (ISSN 1022-1360) is published with 14 volumes per year by WILEY-VCH Verlag GmbH, P. O. Box 10 11 61, 69451 Weinheim, Germany. Air freight and mailing in the USA by Publications Expediting Inc., 200 Meacham Ave., Elmont, NY 11003. Periodicals postage pending at Jamaica, NY 11431. POSTMASTER: send address changes to Macromolecular Symposia, Publications Expediting Inc., 200 Meacham Ave., Elmont, NY 11003.

Printing: Strauss Offsetdruck, Mörlenbach. Binding: J. Schäffer, Grünstadt

Polymer Processing and Structure Relationships Symposium in EUROMAT 2001

held in
Rimini, Italy
June 10–14 2001

Organized by
Associazione Italiana Metallurgia

On behalf of
The Federation of European Material Societies

Symposium Editors

Domenico Acierno
Department of Materials Engineering and Production
University of Naples
Piazzale Tecchio 80
80125 Napoli
Italy

Luciano Di Maio
Department of Chemical and Food Engineering
University of Salerno
Via Ponte don Melillo
84084 Fisciano (SA)
Italy

ISBN 3-527-30470-3

The **tables of contents** of the published issues are displayed on the www.

This service as well as further information on our journals can be found at the following www address:

http://www.wiley-vch.de/home/macrosymp

The service is free for everybody.

Contents of Macromol. Symp. 180

EUROMAT 2001
Rimini, Italy, 2001

* The asterisk indicates the name of the author to whom inquiries should be addressed

Author Index

Preface

In the last few years some significant enhancement in the field of polymer technology has been indubitably observed. In particular the innovations related to synthesis both of new polymers and polymers additives, in combination with new approaches in polymer processing has yielded a very rapid development of polymeric materials with a wide range of structures and related properties. The problem of relating polymeric structure, processing conditions and material properties is becoming, thus, of major importance for both the academic and industrial research. The organization of a symposium dealing with these very complex subjects was mandatory especially in the more general context of a materials-oriented Congress like the EUROMAT. The occurrence of the wide number of polymeric systems and their applications, also represented by several research groups, was evident in a huge variety of papers.

We hope that the selected papers will be of interest for scientists both of polymer chemistry, and polymer blending, processing and recycling. Probably, some of the results are still not included in any industrial application but can stimulate further investigations and will favorably impact research currently performed by the readers.

Having this opportunity, we would like to thank the authors of papers included in this volume for their very valuable contributions and the organizers of the meeting who made presentation of these papers possible.

Domenico Acierno
Luciano Di Maio

Biaxial Orientation of Polyamide Films: Processability and Properties

*L. Di Maio, P. Scarfato, L. Incarnato, D. Acierno**

Department of Chemical and Food Engineering, University of Salerno, Via Ponte don Melillo, 84084 Fisciano, Italy. E-mail: dimaio@dica.unisa.it
*Department of Material and Production Engineering, University of Naples, P.le Tecchio, Naples, Italy

Summary: The production of biaxially oriented films can be performed with two techniques: the tenter frame and the double bubble process. The most important difference between the two processes is that in tubular orientation the MD and TD stretching are simultaneous. This event is very important specially in the case of biaxially oriented polyamide films to avoid hydrogen bonding which occur during the first stretching stage of tenter frame process and cause resistance to subsequent transverse direction orientation (1-4). Double bubble process consists in the stretching of a tubular film where the stretching ratios in the machine and transverse directions are determined by inside bubble pressure and the difference in roll speeds between the two sets of nips rolls that contain the bubble.
The work presented here reports some results about double bubble polyamide films realised with a laboratory scale system. The study concerns the processability (5-6) of two different polyamides: a nylon 6 (PA6) and a copolymer of the nylon 6 with isophthalic acid and isophorondiamine (PA6 – IPA.IPD) at 5% w/w of comonomer. Two main process arrangements where adopted differing for first bubble cooling technique: a water cooling and a air cooling. The films produced in both conditions were then characterised with a particular regard to mechanical and barrier performances. Morphological analyses were also carried out in order to correlate the process conditions to structure and properties of films.

Keywords: POLYAMIDES FILM BLOWING DOUBLE BUBBLE

1 Experimental Part

1.1 Materials

To produce our samples we have used a polyamide 6 (PA6 FL34) and a copolymer of the PA6 with isophthalic acid and isophorondiamine (PA6 – IPA.IPD) at 5% w/w of comonomer, supplied by SNIA (Italy). These materials were chosen as they are widely used for film packaging, due to their good mechanical and barrier performances. The copolymer used here has an improved processability for the blown film process, due to a more branched structure which acts as a bubble stabiliser during the process itself, as was verified in an our previous

 CCC 1022-1360/00/$ 17.50+.50/0

work (7) where the effect of different comonomer structures on properties of copolyamides was shown. Polymer chips, due to their hygroscopicity, were always dried for 12 h at 110°C in a vacuum oven before melt processing.

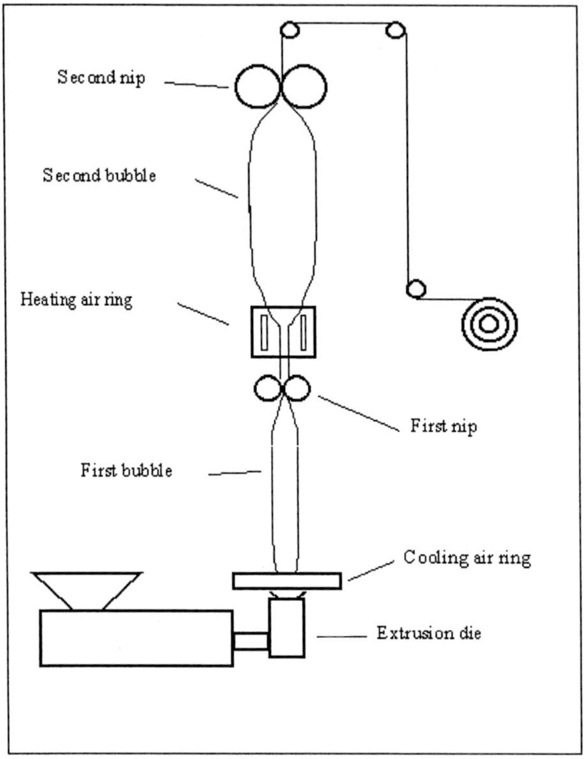

Figure 1. Double bubble apparatus scheme

1.2 Single and Double Bubble Apparatus

Single bubble films were prepared using a GIMAC mod. 12/24 single screw extruder with an anular die of 31 mm for the inner diameter and 32 mm of outer diameter. The single bubble take up frame used was a HAAKE mod. 002-4848.

On the basis of the theory of the film orientation (8-9) we have designed and built in laboratory a double bubble tubular film blowing tower to produce the samples (10). The first phase of our process is a single bubble extrusion that we have carried out using the apparatus described above (HAAKE mod. 002-4848). Over the first pair of nip rolls is positioned an heating apparatus constituted by an air ring, so that we could reach temperature up to 130 degrees. A second pair of nip rolls, that can be opened by two pneumatic pistons, are set over the oven at about the same distance between the first nip rolls and the extrusion die. In fig. 1 can be seen a scheme of the machinery. At the start of the process the film obtained from the single bubble apparatus, is reheated in the air ring and inflated again introducing air at the desired pressure opening the second pair of nip rolls. In this way the pressure of the air entrapped in the second bubble between the first and second pair of nip rolls induces the stretch for the orientation of the molecular chains in the Transverse Direction (TD), while the higher speed of the second pair of nip rolls ensures the orientation in the Machine Direction (MD).

1.3 Process Conditions

The process variables we have used to report the results achieved are, for the single bubble:

Draw Down Ratio: DR = S1nr/Se

where S1nr is the peripheral speed of the nip rolls and Se is the speed of exit of the polymer flow from the extrusion die;

Blow Up Ratio: BR = D1b/Da

where D1b is the diameter of the first bubble and Da is the outer diameter of the annular extrusion die.

For the second bubble:

Second Draw Down Ratio: SDR = S2nr/S1nr

where S1nr and S2nr are the peripheral speeds of the first and second pair of nip rolls;

Second Blow Up Ratio: SBR = D2b/D1b

where D1b and D2b are the diameter of the first and second bubble during the operation.

The condition used to produce film samples are:

Extrusion Temperature:	265°C;
Extrusion screw speed:	10 r.p.m.(corresponding to c.a.12 g/min) ;
DR:	10;
BR:	1;
SDR:	From 1 to 4.5;
SBR:	From 1 to 4;
Blowing up temperature:	From 170 to 230°C (measured at air ring).

1.4 Process Stability

A process stability analysis is reported to evaluate, for each material and operating condition, the range of the draw and blow up extent which allow a proper stable and continuous bubble for a good film production process.

1.5 Film Characterisation

The film samples produced in the conditions above mentioned, were characterised in order to correlate process conditions to film performances. Thermal properties of the film have been determined with a Mettler Differential Scanning Calorimeter DSC30.

Permeability measurements have been carried out with an isostatic gas-permeabilimeter Lyssy GPM 200. The barrier properties were thus evaluated respect to O_2, CO_2 and N_2.

Mechanical properties of the films have been determined with an Instron mod. 4301 testing the films in the two directions, the MD and the TD, with rectangular samples of 11.7 mm width and 110 mm length, the thickness depended on the process conditions while the grip distance was of 50 mm.

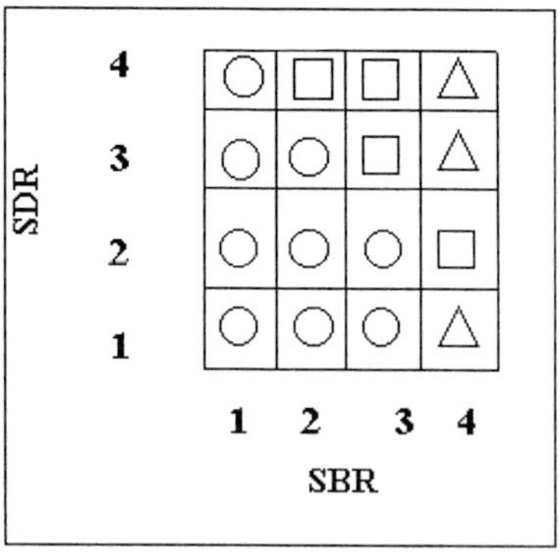

Figure 2. Stability at 170 °C

2 Results and Discussion

2.1 Range of Stability

The production of the samples was performed running various operating conditions in order to check processing limits (from the stability point of view). The parameters modified during the production of the samples were: extrusion rate, periphera l speed of the nip rolls, air pressure inside the first and second bubble. We present the results as functions of SDR and SBR.

The fields of stability for the process with second bubble temperature of 170, 200 and 230 °C are shown in the figures 2-4 on the right respectively from top to bottom. Each geometrical symbol

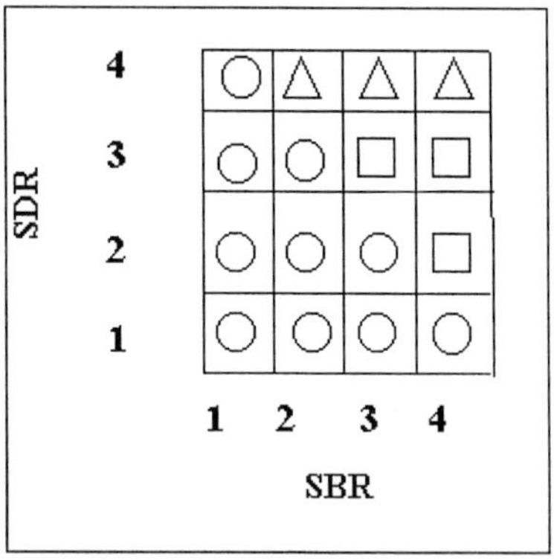

Figure 3. Stability at 200 °C

represents the process stability or instability as follow: O=stable process; =continuous but unstable process; Δ=dis-continuous process. A large number of conditions were tested, but for the sake of brevity here are reported only the most significant. As reported in the figures the bubble stability is strictly dependent on the temperature of the second bubble, which strongly affect also the final properties of the film.

The results have shown that, heating of the second bubble at 200°C, more conditions of stability are possible, in particular higher blow up ratios (SBR) are reached.

In the case of water cooling of first bubble (which ensures a more amorphous material to stretch in the second bubble) the results are similar in terms of SBR and SDR but the temperatures of second bubble can be quite lower. It has to be noted that for the production of double bubble samples it has been necessary to work with a very stable first bubble because every little irregularity in the thickness or in the radius have a negative effect on the blowing and stability of the second bubble. We have thus used a BR = 1 (Blow Ratio of the first bubble) to reach the best results in the blowing of the second bubble so to obtain the best orientation of molecular chains in TD and to ensure the lowest degree of crystallinity of the film from the first bubble at the moment of the blowing of the second bubble.

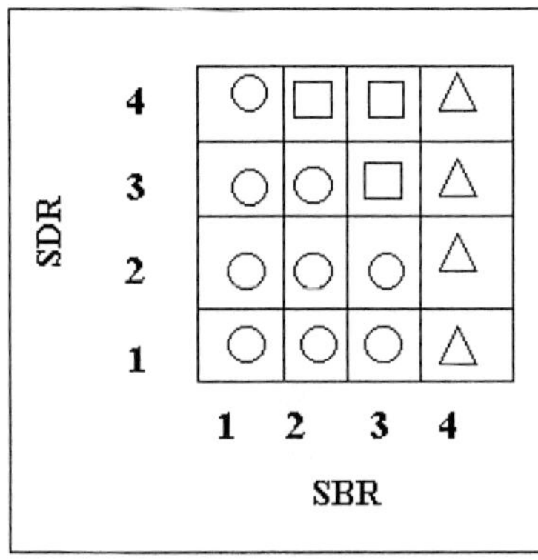

Figure 4. Stability at 230 °C

? ? WAXS Analysis

In order to verify the effectiveness of the process used in producing biaxially oriented films, X-ray diffraction analysis was performed on samples with different SDR and SBR values, obtained from a first bubble completely amorphous and unoriented. In figure 5 is reported as example the WAXD pattern for a film with SBR and SDR both equal to 3 from a second bubble blown up at 200°C. The arrow states the machine direction of the film. The figure clearly

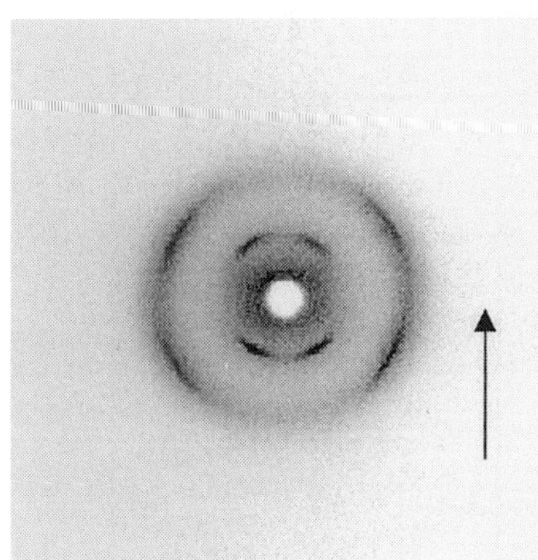

Figure 5. Example of WAXD pattern

shows that this sample is biaxially oriented in MD and TD. Similar results were obtained also analysing film samples having different SDR and SBR values.

2.3 Permeability Tests

Since the gas barrier behaviour of polymeric films is correlated to morphology (orientation, crystallinity) developed during the process, permeability tests were executed for CO_2, O_2 and N_2 with particular regard to operating conditions effect. The results are reported in table 1 for some of the process conditions tested. The permeability is expressed in $cm^3(STP)\cdot cm/cm^2\cdot Pa\cdot s$. Comparing the permeability data measured for different values of stretch ratios and blowing up temperatures, it appears that the lowest permeability results are obtained for the sample E produced at T=200°C. Moreover the SDR and SBR values corresponding to films with lower permeability depend on temperature used for the second bubble. In fact, it is well known (11) that the orientation effect (and so the developed film morphology) of a stretching process is related to the stretching temperature, from which depend the relaxation times of the polymer. As a consequence, both stretching ratios and second bubble blow up temperature have to be optimised to increase the gas barrier performance of the films.

Table 1. Permeability values

Sample	*Stretching ratios*	*Permeability x 10^{14}*			*Second bubble blow up temperature (°C)*
		CO_2	*O_2*	*N_2*	
A	SDR=2 SBR=3	0.94	0.30	0.12	170
B	SDR=2 SBR=2	1.54	0.87	0.62	170
C	SDR=3 SBR=2	0.61	0.19	0.083	170
D	SDR=2 SBR=3	1.04	0.25	0.048	200
E	SDR=2 SBR=2	0.51	0.10	0.024	200
F	SDR=3 SBR=2	2.36	1.26	0.85	200
G	SDR=2 SBR=3	0.53	0.12	0.036	230
H	SDR=2 SBR=2	1.63	1.29	1.00	230
I	SDR=3 SBR=2	1.03	0.27	0.061	230

Mechanical tensile performances of the films were measured both in the machine and transverse directions. As expected, modulus, stress at break and elongation at break are dependent on the extent of film stretching and orientation. In the diagrams below (figures 6 and 7) are reported the modulus and stress at break values for the two stretching directions as a function of the SDR/SBR ratio, which is the measure of the extent of film orientation.

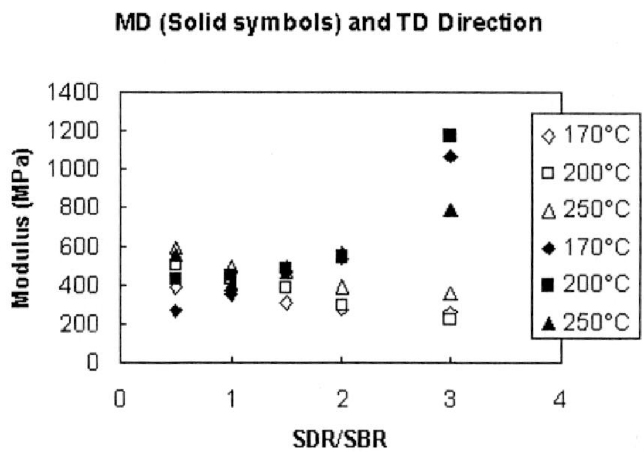

Figure 6. Tensile elastic modulus

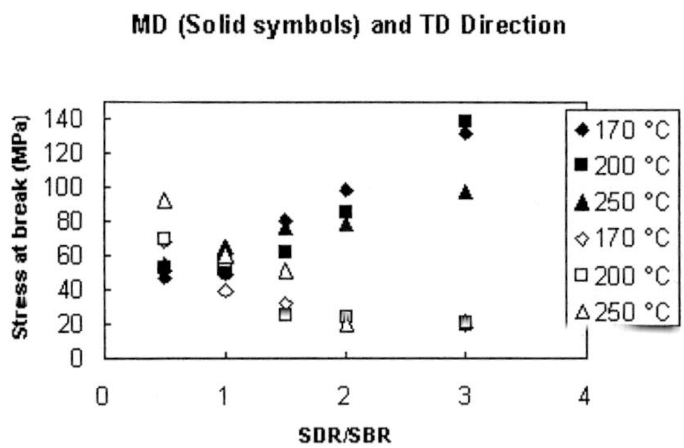

Figure 7. Tensile stress at break

At higher values of this ratio correspond higher modulus and stress at break in the MD. Opposite behaviour was displayed for the same parameters in the TD. Moreover, at SDR/SBR=3, where the differences in the mechanical parameters are consistent, it is evident that also in the case of tensile properties the best performances are shown by film obtained using a temperature of 200°C for blowing up the second bubble.

Conclusions

The production of biaxially oriented PA films was performed in laboratory on an experimental apparatus. The system allowed to test several process conditions and a particular attention was paid to stretching ratios and blowing up temperature of the second bubble. From them, in fact, depend the extent of orientation in MD and TD and the morphology of the samples, that in turn influence the mechanical and barrier properties of the films.

Regarding to the fields of stability for the process, the experiments have shown that, heating the second bubble at 200°C, more conditions of stability are possible and higher SBR values can be reached. Moreover, in terms of performances the films produced at this temperature show also the best barrier properties and mechanical resistance in machine direction. In particular, the film E exhibits the lowest permeability to gases and comparable mechanical resistance values in MD and TD.

References

[1] Kang H.J., White J.L., *Intern. Polymer Processing*, **VII**, p. 38 (1992)
[2] Antti Savolainen, *Polym. Eng. Sci.*, **30**(19) p. 1258 (1990)
[3] Ramesh M.G., *J. Appl. Polym. Sci.*, **48**, p.1635-1648 (1993)
[4] Kanay T., White J.L., *Polymer Science*, **24**, p. 1185 (1984)
[5] Minoshima W., White J.L., *J. Non Newt. Fluid Mech.*, **19**, p. 275 (1986)
[6] Yamane H., White J. L., *Nihon Reoroji Gakkaishi*, **15**, p. 131 (1987)
[7] Scarfato P., Di Maio L., Incarnato L., Acierno D., Mariano A., "Influence of comonomer structure on properties of copolyamide packaging films", *Packag. Tech. Sci.*, submitted (2001)
[8] White J. L., Yamane H., *Pure App. Chem.*, **59** p. 193 (1987)
[9] Abhiraman A.S., *J. Polym. Sci.: Plo. Phy. Ed.*, **21**, p. 583-594 (1983)
[10] Daubeny R.P., Bunn C.W., Brown C.J., *Pro. Roy. Soc.* **A226**, p. 531 (1954)
[11] Van Krevelen D.W., "Properties of Polymers", Elsevier, Amsterdam (1990)

Macromol. Symp. ***180***, *9–22 (2002)*

Binary Blends Based on Poly(vinyl chloride) and Multi-Block Copolymers Containing Poly(ε-caprolactone) and Poly(ethylene glycol) Segments

Maurizio Penco, Luciana Sartore, Fabio Bignotti, Mariapia Rossini, Alberto D'Amore, Federica Fassio*

Dipartimento di Chimica e Fisica per i Materiali e per l'Ingegneria, unità INFM, via Valotti 9, 25123 Brescia Italy

Summary: Binary blends based on poly(vinyl chloride) (PVC) were prepared both by casting from tetrahydrofuran (THF) and by mixing in the melt form, in a discontinuous mixer, PVC and multi-block copolymers containing poly(ε–caprolactone) (PCDT) and poly(ethylene glycol) (PEG) segments. PCDT-PEG copolymers were synthesized using a polycondensation reaction where the α,ω-bis-chloroformate of an oligomeric poly(ε-caprolactone) diol terminated (PCDT) and oligomeric PEG were employed as macromonomers. For comparison purposes, blends PVC with starting oligomers as well as with mixtures containing a typical low molecular plasticizer, dioctylphthalate (DOP), were also prepared. The copolymer miscibility was studied by differential scanning calorimetry (DSC) and FT-IR spectroscopy. The blend morphology was investigated by polarized light microscopy (PLM). A higher miscibility with PVC was observed for copolymers compared to PEG.

Introduction

The supposed toxicity of most commonly used low molecular weight plasticizers employed to produce flexible PVC has required the development of a new generation of low toxicity flexibilizers. This problem is of great importance in food packaging and in biomedical devices, where the materials are in contact with biological fluids or inserted in the human body.

In this connection, an interesting role might be played by polymeric blends containing a low Tg polymer with a high miscibility with PVC [1, 2]. In principle, the

 CCC 1022-1360/00/$ 17.50+.50/0

high molecular weight of the plasticizer can strongly reduce release processes and toxicity.

PVC/poly(ε-caprolactone) (PCL) binary blends are interesting systems because some miscibility exists between the components; furthermore PCL is a non-toxic polymer having a low glass transition temperature (Tg). Partial miscibility has been found with high molecular weight PCL and heterogeneous materials with some amount of semi-crystalline PCL phase have been obtained [3, 4].

Biocompatibility is another important factor for materials used in the biomedical field. The PVC biocompatibility has been increased by surface treatments improving its hydrophilicity [5, 6]. To that purpose, given the well-known biocompatibility of PEG, PVC/poly(ethylene glycol) (PEG) blends might be systems of interest, but unfortunately a low miscibility in general exists between these components [7].

The aim of this work is to study the miscibility of PVC with multi-block copolymers containing PCL and PEG segments. Multi-block copolymers were synthesized by a polycondensation reaction between an oligomeric PEG and the α,ω-bis-chloroformate of oligomeric poly(ε-caprolactone) diol terminated (PCDT) (see Scheme). High molecular weight copolymers can be obtained having PCDT and PEG segments linked through carbonate bonds [8-10]. In principle, the miscibility with PVC can be modified changing the length of PCDT and/or PEG blocks. Therefore, two multi-block copolymers having PCDT segments with $\overline{M}_n$ =1250 and PEG with $\overline{M}_n$ = 1000 and 2000 were synthesised.

PVC blends were prepared using two methods. The former involved casting from THF solutions followed by a thermal treatment. The latter was based on melt mixing, in a discontinuous mixer, using a typical industrial recipe employed to produce serum bags. For comparison purposes, blends containing starting oligomers or dioctylphthalate (DOP) were prepared too. Their miscibility and thermal properties were studied by differential scanning calorimetry (DSC) [12, 13] and FT-IR spectroscopy [14, 15].

Scheme

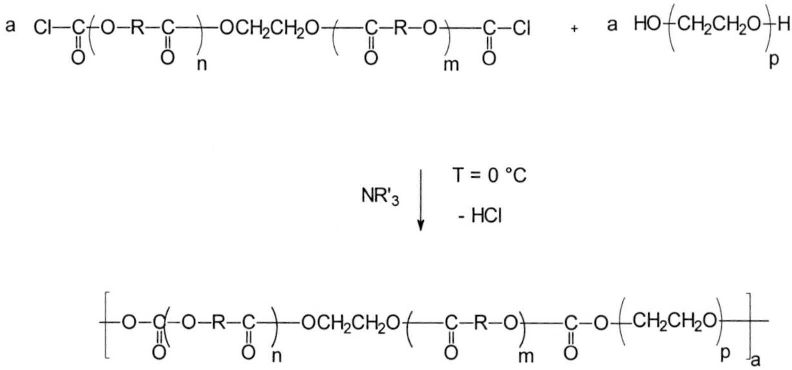

where R= $(CH_2)_5$

Experimental Part

Instruments

The infrared spectra were recorded by a 5300 Jasco FT-IR spectrophotometer on films obtained by casting from solvent or by compression moulding. ^{1}H NMR spectra were run in deuterared chloroform on a 60 MHz Varian 360 A spectrometer, using TMS as internal reference. Viscometric measurements were performed with a Schott Geräte mod AVS 310 semiautomatic viscometer. The GPC experiments were run on an Erma instrument, using Shodex KF columns, operating with chloroform as eluent, flow rate 1 ml/min and a Knauer RI detector. Calibration curves were obtained by poly(ethylene glycol) standards purchased from Polymer Laboratories, Shropshire (U.K.). The DSC analysis was carried out using a Mettler TC11 Thermal Analyser Processor equipped with a Low Temperature Cell DSC30. The DSC curves were obtained by heating a sample of about 10 mg from - 100 to 100°C at 10°C/min under nitrogen flow (10 ml/min).

Methods

Solubility: Solubility experiments were performed by treating the product (0,05 g) with the solvent (1 ml) for 4 h at room temperature. The mixtures were then warmed up to the boiling point.

Intrinsic viscosity. The intrinsic viscosity was evaluated by using an Ubbelohde viscometer at 32°C in chloroform.

Binary blends by casting: Binary blends were prepared from 5 weight % tetrahydrofyran (THF) (b.p. 66 °C) solutions by solvent evaporation at 25 °C and 760 mmHg for a week. The resulting films were treated first at 90 °C for 8 hours and then at 150 °C for 15 minutes, at 1 mmHg of pressure.

Blends by melt mixing: PVC blends were prepared using a Brabender discontinuous mixer employing a typical industrial recipe used for serum bags production. Mixing conditions: 5 minutes at 160 °C, mixing rate 60 rpm.

Starting materials

PVC having $\overline{M}_n$ *=95000* was purchased from Aldrich Co. (Tg = 85.9 °C).

PEG samples of different molecular weights were purchased from Fluka Co., and dried overnight at 40 °C and 0,1 torr before use.

Poly(ε-caprolactone) diol terminated (PCDT) samples having $\overline{M}_n$ =1250 and 2000 were purchased from Aldrich Co. and dried in vacuum at 0.1 torr for 48 hours.

N-ethyldiisopropylamine (99 wt.-% purity) was purchased from Aldrich Co. and dried over calcium hydride for 48 hours.

4-Dimethylaminopyridine (99 wt.-% purity) was purchased from Aldrich Co. and used as received.

Chloroform was a high-grade commercial product purchased from various sources, not containing ethanol as stabilizer. It was allowed to remain some hours over calcium hydride before use.

All other reagents were high-grade commercial products, used without further purification.

Interlite ZX 9004, Lankroflex E2307, Interwax GP 55, Interwax GP 33, Interwax DP 16 were purchased from Akcros Italia.

Synthesis of PCDT1250-PEG1000 segmented copolymer

13 mL (0.025 mol) 20 wt.-% toluene solution of phosgene was placed, under nitrogen atmosphere, in a flask equipped with a stirrer, a gas inlet which could be lifted at various levels, a gas outlet connected with a sodium hydroxide trap and a dropping funnel with pressure-equalizing side arm, into which a mixture of 12.5 g (0.01 mol) dry PCDT having number-average molecular weight 1250 and 3.52 mL (0.02 mol) N-ethyldiisopropylamine (EDIPA) in chloroform (11 mL) had been charged. The flask was cooled by means of an external bath at 5°C and the solution was added dropwise to the phosgene solution under stirring. After addition, stirring at 5°C was continued for further 15 min, then the gas inlet was lowered and nitrogen bubbled through the reaction mixture for 30 min, in order to eliminate the excess phosgene.

To the resulting solution of PCDT bis(chloroformate), a solution of 10.0 g (0.01 mol) PEG 1000 oligomer, 3.52 mL (0.02 mol) EDIPA and 1.23 g (0.01 mol) 4-dimethyl aminopyridine (DMP) in chloroform (35 mL) were added dropwise under nitrogen atmosphere. The reaction mixture was maintained at 5°C for 8 h under stirring, and then for further 10 h while rising to room temperature. After this time, the solvent was eliminated by evaporating under vacuum. The resulting product was then dissolved in chloroform, precipitated with excess isopropanol and immediately washed with ether. The precipitate was dried at room temperature and 0.1 torr (95 wt.-% yield, $\overline{M}_n$ =33900 and $[\eta]$ = 0.38 dL/g).

Synthesis of PCDT1250-PEG2000 segmented copolymer

PCDT1250-PEG2000 segmented copolymer was prepared in the same way as PCDT1250-PEG1000 by substituting PEG having number-average molecular weight 2000 (20 g, 0.01 mol) for PEG 1000. Yield of 95 wt.-% ($\overline{M}_n$ =63500; $[\eta]$ = 0.52 dL/g) .

Characterisations of PCDT1250-PEG1000

IR spectra: the IR spectrum of PCDT1250-PEG1000 showed bands at: 2945-2868 cm^{-1} (ν, CH); 1726 cm^{-1} (ν, C=O); 1300-1000 cm^{-1} (ν, C-C(=O)-O; ν, C-C-O). The IR spectra of PCDT1250-PEG2000, though on the whole qualitatively similar, showed an increase of the relative intensity of the band at 1150-1085 cm^{-1} (ν, C-O-C); the ester band (ν, C=O) is shifted at 1732 cm^{-1}.

^{1}H-NMR spectra: the ^{1}H-NMR($CDCl_3$) spectra of PCDT-PEG products showed similar peaks but obviously with different relative intensities. In particular: δ =3.9 – 4.3 (m, CH_2OC=O), δ =3,6 (s; CH_2OCH_2); δ =1.2- 1.9 (m; CH_2-CH_2).

Results and Discussion

1. PCDT/PEG multi-block copolymers

The synthetic process for the preparation of PCDT-PEG segmented copolymers involves the activation of PCDT oligomers by means of phosgene, resulting in the introduction of terminal chloroformate groups. Without isolating, activated PCDTs are then reacted with PEG oligomers, as shown in the scheme. In this way multi-block copolymers having segments linked with carbonate bonds are obtained. We reported this reaction in previous papers dealing with the synthesis of other segmented copolymers based on PLGA. In that case copolymers having a poly(ester-carbonate) structure resulted [8-10].

The molecular weight of oligomeric PEGs employed was selected using a simple miscibility prediction scheme based on the solubility parameters [16, 17]. Values of $[\chi_{AB} - (\chi_{AB})_{cr}]$ *versus* the $\overline{M}_n$ of starting oligomers for blends with PVC ($\overline{M}_n$ = 95000) are shown in Figure 1. Miscibility is expected when this parameter is negative or it equals zero (approximately for $\overline{M}_n \leq 1000$). On the other hand it is well-known that good miscibility can exist in PVC/PCDT blends [18, 19], in agreement with the results of the prediction scheme adopted (Figure 1).

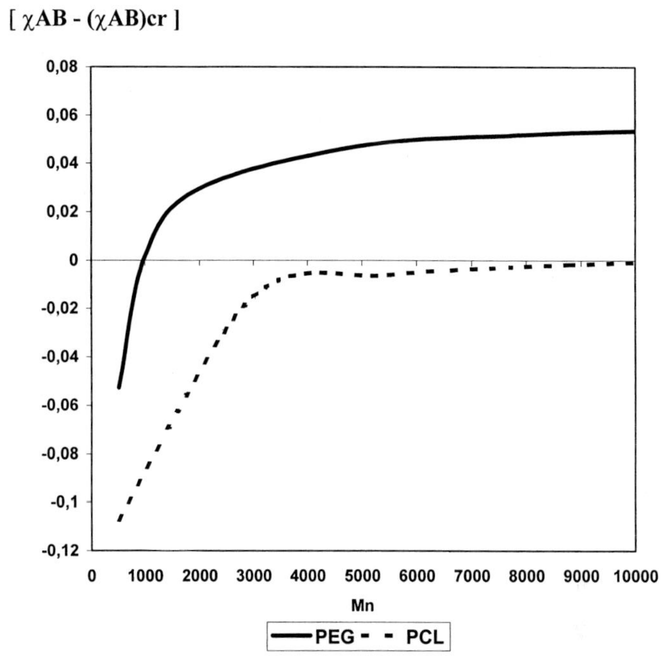

Figure 1: [χ_{AB} - $(\chi_{AB})_{cr}$] *versus* $\overline{M}_n$ of starting oligomers for blends with PVC with $\overline{M}_n$ 95000.

Therefore the miscibility of PCDT-PEG copolymers is expected to be considerably influenced by the length of the PEG segments and by the PEG content. On that basis, PEGs with $\overline{M}_n$ 1000 and 2000 were selected. The choice of the $\overline{M}_n$ values for the PCDT oligomers (1250 and 2000) was dictated by their low Tg (-77.7 °C and –60.2 °C, respectively).

The composition, intrinsic viscosity, and thermal properties of the copolymers investigated are reported in Table 1, along with calorimetric data of starting oligomers. It should be noticed that PCDT-PEG copolymers show a single Tg whose value is higher than in starting PCDT.

Table 1: Molecular weight characterisation and DSC data of starting oligomers and block-copolymers

Polymer	Molecular weight Characterization				Calorimetric data		
	[η] dL/g-	a) $x10^{-3}$	a) $x10^{-3}$	D b)	Tg c) (°C)	Tm c) (°C)	ΔH c) (j/g)
PCDT 1250	0.09	1.25	-	-	-77.7	58.3	74.9
PEG 1000	-0.03	1.00	-	-	-	38.3	127.0
PEG 2000	-0.09	2.00	-	-	-	56.1	190.0
PCDT1250-PEG1000	0.38	33.90	64.4	1.9	-69.5	45.5	51.8
PCDT1250-PEG2000	0.52	63.50	127.0	2.0	-56.8	40.6	62.4

a) Calculated by SEC using a PEG calibration curve.
b) Polydispersity index: $\overline{M}_w / \overline{M}_n$.
c) under nitrogen (scanning rate: 10°C/min)

Furthermore, a wide fusion peek located in the same temperature range as in PEG and PCDT and having an enthalpy of fusion substantially lower than in starting oligomers was also detected. These data would suggest that some miscibility exists between the PCDT and PEG segments in the above copolymers.

2. PVC blends

Binary blends obtained by casting

Binary blends were prepared by solvent evaporation from THF solutions having a 5 weight % polymer content. The samples were treated first at 90 °C for 8 hours and then at 150 °C for 15 minutes, under vacuum. The composition and thermal properties of the blends prepared are summarized in Table 2. The percentage of crystallinity of PEG in the blends in comparison with that of pure oligomer was evaluated according to the equation: $C = \Delta H_m \cdot 100 / w \cdot \Delta H^{\circ}_m$; where ΔH°_m is the heat of fusion of starting oligomers or copolymers, w is their weight fraction in the blend and ΔH_m is the heat of fusion of the blend.

Table 2: DSC data of PVC blends obtained by casting

Polymer		Calorimetric data[a)]				
Type	wt-%	Tg (°C)	Tm (°C)	ΔH (j/g)	C[b)] (%)	ΔTm[c)] (°C)
PVC	-	81.0	-	-	-	-
PEG 1000	10	48.7	28.7	1.4	11	9.6
PEG 1000	20	44.5	35.0	1.9	8	3.3
PEG 1000	30	44.0	33.6	19.3	51	4.7
PEG 2000	10	42.5	47.5	1.6	8	8.6
PEG 2000	20	-	49.3	15.2	40	6.8
PEG 2000	30	-	50.9	28.4	50	5.1
PCDT 1250	10	39.0	-	-	-	-
PCDT 1250	20	-8.5	-	-	-	-
PCDT 1250	30	-23.5	-	-	-	-
PCDT-PEG1000	20	11.0	-	-	-	-
PCDT-PEG2000	20	-33.0	42.6	2.0	16	-2.0

a) under nitrogen (scanning rate: 10°C/min)
b) Percentage of crystallinity evaluated by means of the following equation: $C = \Delta H_m \cdot 100 / w \cdot \Delta H^{\circ}_m$ where ΔH°_m is the heat of fusion of starting oligomers or copolymers, ΔH_m is the heat of fusion of the blend and w is the their weight fraction in the blend.
c) melting temperature depression: Tm° - Tm , where Tm° and Tm are the melting point of pure crystallisable component and of blend respectively.

The calorimetric data of PVC/PEG blends show a very limited miscibility. Even if all materials contain a crystalline PEG phase, these data suggest that PEG1000 is more miscible with PVC than PEG2000. In fact, crystallinity values of 11%, 8%, 51 %, and 8%, 40%, 50% were calculated for blends containing 10, 20 and 30 weight-% of PEG1000 and PEG2000, respectively. On the contrary, only a limited difference exists between the melting depression temperature (ΔTm) of these two series. For instance ΔTm is 9.6°C and 8.6°C for PVC/PEG1000 and PVC/PEG2000 with a 10 weight-% oligomer content, respectively. An anomalous value of ΔTm is exhibited by PVC/PEG1000 (20 weight-%) as a ΔTm near to that of PVC/PEG2000 is expected. Tg's in the range of 40 ÷ 50°C were in general detected. FT-IR spectroscopy confirms the existence of some degree of interaction between PVC and PEG1000; in fact, the ether band is shifted from 1109 cm^{-1} (pure PEG 1000) to 1103 cm^{-1} (20 weight-% PVC/PEG1000).

On the other hand, full miscibility exists in PVC/PCDT1250 blends within the range of compositions examined. All blends are amorphous materials showing a single Tg, the value of which decreases with the PCDT content from 40°C to 25°C. FT-IR spectroscopy shows a shift of the ester band from 1724 cm^{-1} (pure PCDT1250) to 1732 cm^{-1} (10 weight-% PVC/PCDT1250). This shifting is ascribed to hydrogen bond interactions occurring between PVC and PCDT.

The miscibility of PCDT-PEG copolymers with PVC is strongly dependent on the length of the PEG segments, *i.e.* on their weight fraction. In fact, the PVC/PCDT-PEG1000 blend does not show appreciable crystallinity and displays a single Tg at 11 °C, a value substantially lower than in the PVC/PCDT1250 blend containing the same amount of PCDT (10 weight-% of PCDT1250, Tg = 39 °C). In contrast, a peek of fusion appears in PVC/PCDT-PEG2000. The lower melting point (Tm = 42.6 °C) compared to the PVC/PEG2000 blend having a comparable PEG2000 content (10 weight-%, Tm = 47.5 °C) suggests that some miscibility exists. The morphological analysis, performed by polarized light microscopy, confirms that the PEG2000 miscibility is increased if a multi-block copolymer is employed (see Figure 2.a and 2.b). The FT-IR analysis shows a shift of the ester and ether bands, as observed in PVC/PCDT and PVC/PEG blends.

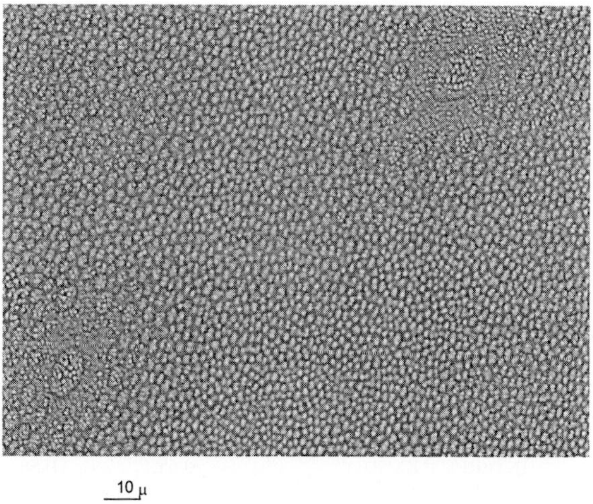

(a)

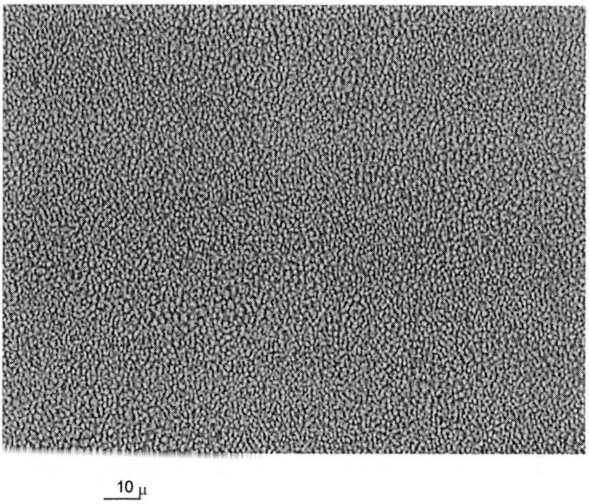

(b)

Figure 2: Polarized light microscopy (PLM) of PVC/PEG2000 10 wt.-% (a) and PVC/PCDT-PEG2000 20 wt.-% blends.

Blends by mixing melt

PVC blends were prepared using a Brabender discontinuous mixer at 160 °C employing a typical industrial recipe used for serum bags production (see Table 3).

Table 3: Typical recipe used in the discontinuous mixer

Product	weight (g)	
PVC	34.00	28.00
Plasticizer	6.00	12.00
INTERLITE ZX 9004[a)] (Ca/Zn stabilizer)	0.68	0.56
LANKROFLEX E2307[a)] (Epoxidized soya bean oil)	1.36	1.12
INTERWAX GP 55[a)] (Lubricant)	0.68	0.56
INTERWAX GP 33V[a)] (Lubricant)	0.34	0.28
INTERWAX GP 16[a)] (Lubricant)	0.34	0.28

a) Products purchased from AKCROS CHEMICALS

In particular, mixtures containing oligomeric PCDT 2000 and DOP were prepared for comparison with PVC/PCDT-PEG. The thermal properties of blends, evaluated by means of DSC, are summarized in Table 4. A high miscibility is noticed in systems containing oligomeric PCDT. In fact, the blends exhibit optical clarity and only a single Tg whose value strongly depends on the composition. For instance, a Tg of 6.6°C and -22.2°C is observed for the 15 and 30 weight-% content of oligomer, respectively. It should be noticed that the latter Tg is very close to that of the PVC/DOP blend with the same amount of plasticizer (Tg = -20.8). No melting peeks are observed in these materials.

Only a single Tg is detected in blends containing PCDT-PEG1000 (*i.e.* –11 °C in the sample at 30 %). Also, no appreciable crystallinity is present and optical clarity is observed. Therefore, it is possible to say that a high miscibility is maintained even when typical additives are added.

In the blend containing PCDT-PEG2000 a lower miscibility is observed. In fact, the DSC traces show the presence of a crystalline phase probably related to the PEG blocks (Tm = 38.1°C).

Table 4: PVC blends obtained in Brabender mixer

Plasticizer		**Calorimetric data** [a)]		
Type	**Wt.-%**	**Tg (°C)**	**Tm (°C)**	**ΔH (j/g)**
DOP	15	-9.1	-	-
DOP	30	-20.8	-	-
PCDT 2000	15	6.6	-	-
PCDT 2000	30	-22.2	-	-
PCDT1250-PEG1000	30	-11.6	-	-
PCDT1250-PEG2000	30	-30.9	38.1	1.7

a) under nitrogen (scanning rate: 10°C/min)

Conclusions

Binary blends based on PVC were prepared by casting using multi-block copolymers containing PCDT and PEG segments. The synthetic scheme employed gives the opportunity of tailoring the miscibility between the copolymers and PVC. In fact, the presence of the PCDT blocks increases the miscibility of the PEG segments. A process of melt mixing, employing typical calcium/zinc stabilizers, leads to results in agreement with those obtained by casting.

[1] D. Braun, M. Bergmann, *Angew. Makromol.Chem.*, **1992**, 199, 191.
[2] J. R. Pena, M. Hidalgo, C. Mijangos, *J. Appl. Polym. Sci.*, **2000**, 75, 1303.
[3] S. Y. KwaK, *J. Appl. Polym. Sci.*, **1994**, 53, 1823.
[4] C. Chiu, K. Min, *Polym. Int.*, **2000**, 49, 223.
[5] Y. Ishikawa, S. Sasakawa, *Radiation Physicsand Chemistry, International Journal of Radiation, Applications and Instrumentation Part C*, **1992**, 39, 561.
[6] D. Klee, R. V. Villari, H. Hocker, B. Dekker, C. Mittermayer, *J. Mater. Sci., Mater. Med.,* **1994**, 5, 592.
[7] C. Marco, M. A. Gomez, J. G. Fatou, A. Etxeberria, M. M. Elorza, J. J. Iruin, *Eur. Polym. J.*, **1993**, 29, 1477.
[8] M. Penco, E. Ranucci, F. Bignotti, P. Ferruti, *Macromol. Rapid Commun.*, **1994**, 15, 683.
[9] M. Penco, M. Becattini, P. Ferruti, S. D'antone, R. Deghenghi, *Polym. Adv. Technol.*, **1996**, 7, 536.
[10] M. Penco, R. Donetti, P. Ferruti, *Macromol. Chem. Phys.*, **1998**, 199, 1737.
[11] C. Prasharn, A. M. Jamieson, *Polymer*, **1997**, 38, 1273.
[12] R. E. Zozaya, M. Buso, G. Pezzin, *Journal of Vinyl Technology*, **1993**, 15, 132.
[13] S. Shen, J. M. Torkelson, *Macromolecules*, **1992**, 25, 721.
[14] G. Ruggeri, R. Bertani, M. Aglietto, A. Dalessio, E. Benedetti, *Polym. Int.*, **1994**, 34, 1.
[15] D. Braun, N. Eidam, D. Leiss, *Makromol. Chem., Macromol. Symp.,* **1991**, 52, 105.
[16] D. R. Paul, S. Newman, *Polymer Blends*, vol. 1, Academic Press, New York 1978, p.45.
[17] J. C. Huang, D. C. Chang, R. D. Deanin, *Advances in Polymer Technology*, **1993**, 12, 81.
[18] J. L. Brode, J. V. Koleske, *J. Macromol. Sci., Chem.*, **1972**, A6, 1109.
[19] J. V. Koleske, R. D. Lundberg, *J. Polym. Sci.Part A-2,* , **1969**, 7, 795.

Cast-Extruded Syndiotactic Polypropylene Films: Preliminary Structural and Mechanical Results

Liberata Guadagno[*], *Carlo Naddeo, Concetta D'Aniello, Luciano Di Maio, Vittoria Vittoria, Domenico Acierno*[§]

Department of Chemical and Food Engineering
University of Salerno
Via Ponte Don Melillo 84084 Fisciano - SA, Italy

[§] Department of Materials and Production Engineeering
University of Naples "Federico II" – Piazzale Tecchio 80100 Napoli, Italy

Summary: Syndiotactic polypropylene (sPP) was cast-extruded with a laboratory single screw extruder, obtaining a crystalline and still highly transparent film.
The structural studies showed that the film crystallized at room temperature in the disordered helical form I, containing a fraction of a mesophase with the chains in trans-planar conformation. X-ray patterns, taken either along the extrusion direction (MD) or along the orthogonal directions (TD and ND), indicated a low orientation of the *c* axis parallel to the machine direction (MD) and a partial orientation of the *a* axis along the transverse direction (TD). Mechanical properties performed in either direction showed a very similar behaviour, but a different strain at the breaking point. The mechanical parameters were derived in both directions.

Introduction

The use of new metallocene catalysts in recent years has allowed a very rapid development of polymers with a wide range of structures and related properties,

[*] Corresponding author: Dipartimento di Ingegneria Chimica e Alimentare - Università di Salerno - via Ponte don Melillo, 84084 Fisciano (SA), Italy. *E-mail address: lguadagno*@unisa.it

 CCC 1022-1360/00/$ 17.50+.50/0

depending on the processing conditions. Among them syndiotactic polypropylene (sPP), recently obtained with high tacticity and high molecular weight[1-3], is receiving a great attention owing to its new properties respect to the isotactic isomer. In the case of sPP the difficulty of correlating the physical properties to the structural organization, as determined by the processing procedures, is principally due to its very complex polymorphic behavior, not yet fully clarified[4-15].

Four crystalline forms of sPP have been described so far, characterized by either different chain conformation or crystalline packing. Form I and form II are characterized by chains in helical conformation[4-7], whereas form III and form IV present chains in trans-planar and $(T_6G_2T_2G_2)_n$ conformations, respectively[8-9]. Furthermore, depending on degree of stereo-regularity and on the mechanical and thermal history of the sample, different amounts of disorder in the crystalline packing can be evidenced[16-17].

Beside the crystalline phase structure, and the presence of disordered phases, an important parameter to be taken into account is the orientation state of either the crystalline or the amorphous phase. This parameter largely dictates the performance of semi-crystalline samples, processed in conditions in which orientation can develop[18-21].

In this paper we present some preliminary experimental results about the structure and the mechanical properties of sPP films obtained by "cast-extrusion" process. This is an industrial process in which some orientation and better properties can be achieved.

Experimental

Syndiotactic polypropylene (sPP) was cast-extruded with a laboratory single screw extruder at 200°C into a film 30 μm thick (sample CE).

Fiber X-ray diffractions spectra were obtained using a PHILIPS PW 1710 Powder Diffractometer (Ni-filtered Cu K_α radiation) operating under vacuum by means of a cylindrical camera with a radius of 57.3 mm. The imaging plate FUJI BAS-1800 system was used to record the diffraction patterns.

The infrared spectra were obtained in absorbance by using a FT-IR BRUKER IFS66 spectrophotometer with a resolution of 4 cm^{-1} (32 scans collected).

Differential scanning calorimetry (DSC) results were obtained using a METTLER TA 4000-DSC 30 thermoanalyzer purged with nitrogen and chilled with liquid nitrogen for sub-ambient measurements. Temperature range was –100°C/200°C at a heating rate of 20°C/min.
Birefringence results were obtained using a polarizing optical microscope JENAPOL equipped with Eringhause compensator.
Mechanical tensile properties were analyzed with an Instron 4301 Dynamometer. Samples of film of 120 x 10 mm x mm were cut in both machine and transverse direction. Cross head speed and grip distance were set at 50 mm/min and 100 mm respectively.

Results and Discussion

Structural Organization

In some cases, the cast-extrusion can produce a film with an uniplanar orientation: to investigate this possibility, the X-ray patterns have to be taken along the extrusion direction (machine direction, MD) and along the orthogonal directions (through direction ND and transverse direction TD). The X-ray diffraction pattern of the cast-extruded sample (CE) obtained along the MD direction shows only Debye-Sherrer rings. In Fig.1 we show the X-ray diffraction pattern of the cast extruded sample (CE) taken along the ND direction. The same pattern is obtained along TD direction.
All the reflections, the spacing and the diffraction angles 2ϑ correspond to the disordered form I. In both patterns, Debye-Sherrer rings from (200) reciprocal plane are the most prominent, showing a maximum of intensity along the equatorial plane. As a matter of fact, only the (200) reflection of the helical form I, corresponding to 12.3° of 2ϑ is polarized along the equator, indicating a partial orientation of the *a* axis along the transverse direction (TD) perpendicular to the machine direction (MD). The second reflection, indexed as 010 of form I and corresponding to 15.9° of 2ϑ is not polarized, whereas the third reflection corresponding to 20.8° of 2ϑ and indexed as (111) of form I shows a very low polarization along the first layer, corresponding to the helical identity

period of 7.45 Å. This indicates a low orientation of the *c* axis parallel to the machine direction.

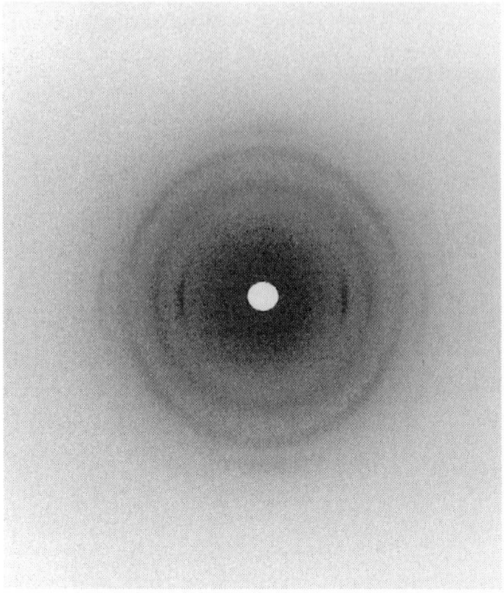

Fig 1. X-ray diffraction pattern of sample CE along ND direction

It is worth noting that the second reflection, at 15.9° of 2ϑ, is not sharp, showing a diffuse halo at 17° of 2ϑ. This reflection, already found in samples crystallized at low temperatures[22-25], was attributed to a trans-planar mesophase, containing lateral disorder in the packing of the chains. This phase was probably formed, in a small fraction, during the rapid cooling of the cast-extruded film at room temperature.

To confirm the partial presence of chains in trans-planar conformation, in Fig.2 we show the FTIR spectrum in absorbance (1300-700 cm^{-1}) of sample CE.

Infrared analysis is very sensitive for the chain conformation determination: as a matter of fact helical and trans planar bands have been evidenced since the first preparation of the syndiotactic isomer of polypropylene[26-27].

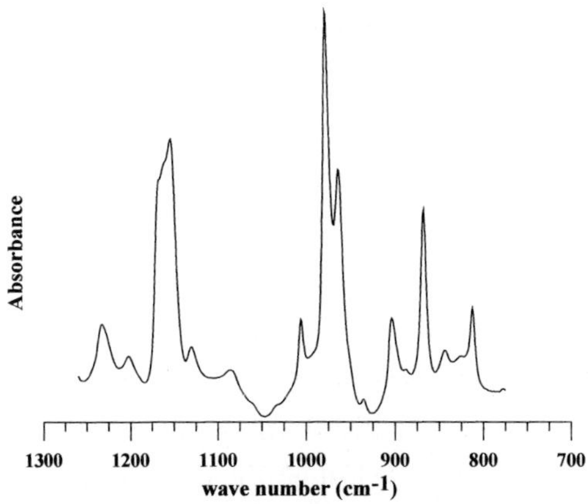

Fig 2. Infrared spectrum of sample CE

We observe that the bands of the helical form of sPP, appearing at 810, 868, 977, and 1005 cm^{-1}, are present and well developed, but the trans-planar bands at 831, 963, and 1132 cm^{-1} are evident too, although of reduced intensity, indicating that a small fraction of chains organized in a different conformation. Therefore, during the extrusion process, the film crystallizes at room temperature prevalently in the helical form I, and still containing a fraction of mesophase with the chains in trans-planar conformation.

In Fig.3 the thermogram obtained in the differential scanning calorimeter (DSC) is reported.

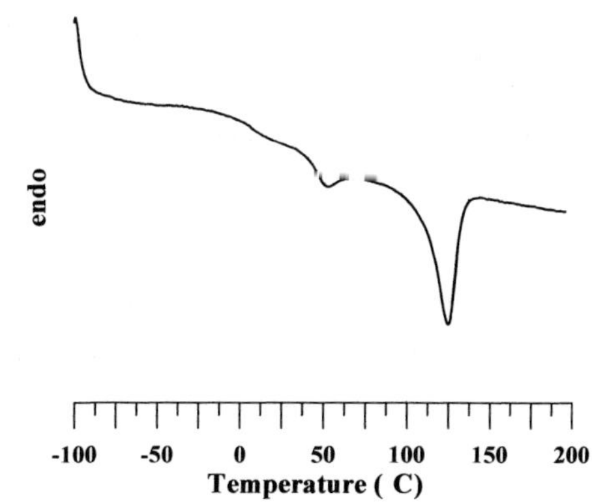

Fig 3. d.s.c. thermogram of sample CE

The curve shows the melting peak with the maximum at 124.5°C, and a melting enthalpy, ΔH_m = 30.5 J/g. This value corresponds to a crystallinity of 16% , obtained using for ΔH_o the value of 8 kJ/mol[28]. At lower temperatures we observe the deflection from the baseline, due to the glass transition, appearing at about 5.2°C. Afterwards, a second deflection, appearing as a small endothermic peak, centered at 52.8°C, is evident in the curve. It was already associated to the melting of the mesophase[23-25] : also in this case the X-ray diffractograms as well as the FTIR spectra suggest that there is in the sample a fraction of material neither in the helical crystalline nor in the amorphous state.
The orientation of the cast extruded film was investigated by birefringence measurements, performed on different strips cut along the film width. This procedure was followed because in the cast extrusion process a different orientation can be achieved in the central part of the film, respect to the lateral strips.
In Fig.4 we show the birefringence, Δn, as a function of the distance from the center of the film.

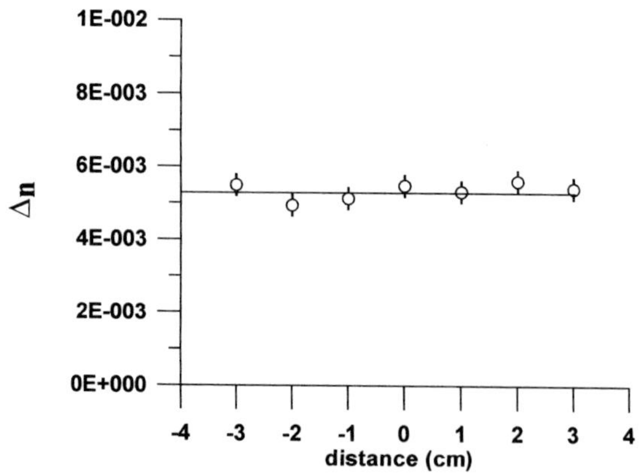

Fig 4. Birifrangence index along the width of the film

We can observe that the experimental points fit the same straight line, whose mean value is 5.3 x 10^{-3}. This result indicates that the orientation is not very high, as already shown by the X-ray patterns, and it is uniformly distributed along the width of the film.

Mechanical Properties

Representative stress-strain curves, performed both in the machine direction (MD) and in the transverse direction (TD), are shown in Fig.5.

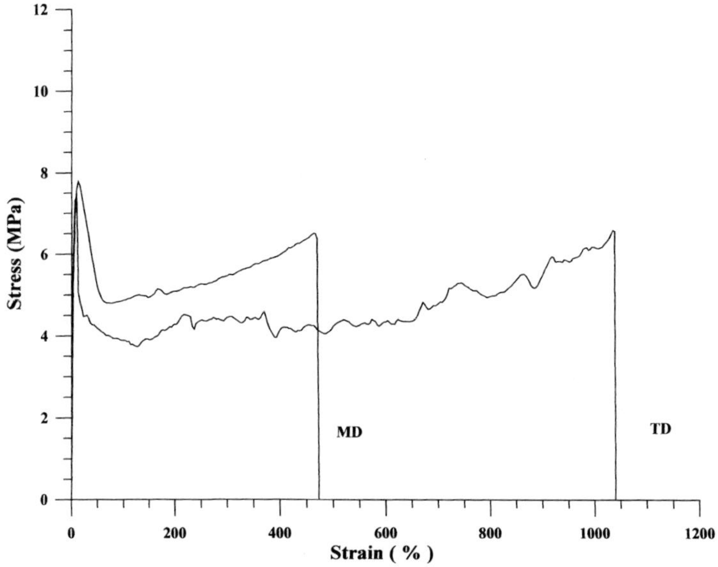

Fig 5. Sress-Strain curves along MD and TD directions

In Table 1 we report the average mechanical parameters of the material. We can observe that almost all the mechanical parameters are very similar in either direction. In particular the elastic moduli are the same, in the experimental error, and the yield stress is very near in the two directions. Only the strain at break is very different, being 5% for sample CE-MD and 11% in the transverse direction (sample CE-TD), indicating that the chain arrangement is different in the two directions.

In spite of this large difference, the stress of breaking is just the same, that is 6.4 MPa.

Table 1

The elastic modulus, E (Mpa), the yield stress σ_y (Mpa), the breaking stress σ_b (Mpa) and the draw ratio of breaking λ_b for sample CE tested along the machine direction (CE-MD) and in the transverse direction (CE-TD).

Sample	E (MPa)	σ_y (MPa)	σ_b (MPa)	λ_b
CE-MD	234	7.8	6.4	5
CE-TD	228	7.5	6.4	11

This work was supported by Ministero dell'Università e della Ricerca Scientifica e Tecnologica MURST (PRIN 2000 "*Catalizzatori di coordinazione e polimeri a struttura mirata*").

References

[1] Ewen, J.A.; Jones, J.A.; Razavi, A.; Ferrara, J. D.; *J. Am. Soc.* **1988**, *101*, 6255.
[2] Longo, P.; Proto, A.; Grassi, A.; Ammendola, P.; *Macromolecules* **1992**, *24*, 462.
[3] Grisi, F.; Longo, P.; Zambelli, A.; Ewen, J. A.; *J. Mol. Cat. A: Chemical* **1999**, *140*, 225.
[4] Lotz, B; Lovinger, A. J.; Cais, R. E.; *Macromolecules* **1988,** *21*, 2375.
[5] Lovinger A.J.; Lotz, B.; Cais, R.E. ; *Polymer* **1990**, *31, 2253.*
[6] Lovinger, A. J.; Davis, D.D.; Lotz, B.; *Macromolecules* **1991,** *24, 552.*
[7] Lovinger, A.J.; Lotz, B.; Davis, D.D.; Padden, F.J.; *Macromolecules* **1993,** *26, 3494.*
[8] Chatani, Y. ; Maruyama, H.; Noguchi, K.; Asanuma, T.; Shiomura, T.; *J Polym. Sci. Part C* **1990**, *28*, 393.
[9] Chatani, Y. ; Maruyama, H.; Asanuma, T.; Shiomura, T.J. ; *J. Plym. Sci. Poly. Phys. Ed.* **1991,** *29*, 1649.
[10] De Rosa, C. ; Corradini, P.; *Macromolecules* **1993**, *26*, 5711.
[11] De Rosa, C.; Auriemma, F.; Vinti, V.; *Macromolecules* **1997,** *30*, 4137.
[12] De Rosa, C.; Auriemma, F.; Vinti, V.; *Macromolecules* **1998,** *31*, 7450.
[13] Lacks, D.J.; *Macromolecules* **1996,** *29*, 1849.
[14] Palmo, K.; Krimm, S.; *Macromolecules* **1996**, *29*, 8549.
[15] Nakaoki, T.; Ohira, Y.; Hayashi, H.; *Macromolecules* **1998**, *31*, 2705.
[16] Auriemma, F.; Lewis, R.H. ; Spiess, H. W.; De Rosa, C. ; *Macromol. Chem. Phys.* **1995**, *196*, 4011.
[17] De Rosa, C.; Auriemma, F.; Corradini, P.; *Makcromol. Chem.* **1996**, *29*, 7452.
[18] Uehara, H.; Yamazaki, Y.; Kanamoto, T.; *Polymer* **1996,** *37*, 57.
[19] Loos, J.; Buhk, M.; Petermann, J.; Zoumis, K.; Kaminsky, W.; *Polymer* **1996,** *37*, 387.
[20] Loos, J.; Petermann, J.; Waldofner, A.; *Colloid Polymer Sci.* **1997**, *275*, 1088.
[21] D'Aniello, C.; Guadagno, L.; Naddeo,C.; Vittoria,V.; *Macromolecular Rapid Commun.* **2001**, *22*, 104
[22] Vittoria, V.; Guadagno, L.; Comotti, A.; Simonutti, R.; Auriemma, F.; De Rosa, C.; *Macromolecules,* **2000**, *33*, 6200.
[23] Guadagno, L.; Fontanella, C.; Vittoria, V.; Longo, P.; *J. Polym. Sci., Part C* **1999,** *37*, 173
[24] Guadagno, L; D'Aniello, C; Naddeo, C.; Vittoria, V; *Macromolecules,* **2000,** *33*, 6023.
[25] L. Guadagno; L. D'Arienzo; V. Vittoria; P. Longo; G. Romano; **2000**, *J. Macromol. Sci. Phys. Ed., B39,* 425
[26] Corradini, P.; Natta, G.; Ganis, P.; Temussi, P.; *J. Polym. Sci.C,* **1967,** *16*, 2477.
[27] Natta, G.; Corradini, P.; Ganis, P.; *Makromol. Chem.* **1960,** *39*, 238.
[28] Rodriguez-Arnold, J.; Zhang, A.; Cheng, S.Z.D.; Lovinger, A.J.; *Polymer*, **1994**, 25, 1884.

Macromol. Symp. ***180,*** *33–41 (2002)*

Gamma Radiation of EVA-AA/MMA Swollen Systems to Obtain Mechanically Improved Blends

G. Spadaro[1], *C. Dispenza*[1], *A. M. Visco*[2], *A. Valenza*[2]

[1] Dipartimento di Ingegneria Chimica dei Processi e dei Materiali University of Palermo, Viale delle Scienze, 90128 Palermo, Italy
[2] Dipartimento di Chimica Industriale e Ingegneria dei Materiali, University of Messina, Salita Sperone 31, 98166 S. Agata, Messina, Italy

Summary: In this work a "new" way to obtain "compatibilised" blends of rubber-glassy thermoplastic components is presented. Polyethylen-vinyl acetate acrylic acid terpolymer is swollen by methylmetacrylate and the system is irradiated through gamma rays in order to induce polymerisation of the monomer. Solubility and dynamic mechanical tests indicate the formation of a two phase system, even though in presence of strong interactions between the two components. Mechanical tensile behaviour is that of a ductile material with appreciable modulus values.

Introduction

Blending of elastomers with brittle termoplastics can be considered a way to produce "new" materials with improved mechanical behaviour. In designing these materials the main problems arise from their thermodinamic incompatibility, better performances can be obtained inducing interactions between the polymer components, giving rise to the formation of "compatibilised" blends. Several compatibilisation procedures are reported in literature [1-4].

It is known that the interaction of ionising radiation with matter causes the formation, through intermediate exited species, of neutral and charged free radicals, which can undergo to further evolution with significant modification of the final molecular structure of the irradiated material [5, 6].

In previous papers[7,8] the use of both gamma rays and elctron beam to produce polymethylmetacrylate-rubber "compatibilised" blends has been suggested. The rubber was dissolved in methylmetacrylate and the monomer was polymerised by

 CCC 1022-1360/00/$ 17.50+.50/0

ionising radiation. During polymerisation rubber phase separation occurred, and the presence of strong interactions between the two phases cuased significant improvement of some mechanical properties.

In this work a new way to improve the thoughness of polymethylmetacrylate or, in turn, the stiffness of a rubber, is presented. Methylmetacrylate is swollen in ethylen-vinyl acetate-acrylic acid terpolymer and polymerised through gamma rays. Solubility tests, mechanical tensile and dynamic thermal mechanical analysis show the presence of strong interactions between the two polymer components.

Experimental

Materials

- Methylmetacrylate (MMA) inhibited with 100 ppm of hydroquinone (Merck). MMA was freed from inhibitor by distillation under the pressure of 1 Pa at about 40°C.
- Ethylene-vinylacetate-acrylic acid terpolymer (EVA-AA), ELVAX 4355 by Dupont, containing the 25% of vinyl acetate and 0.8% of acrylic acid (MFI=6.0).

Sample preparation

Sheets of EVA-AA rubber were swollen in methylmethacrylate under vacuum for 24 h. Before irradiation samples were put in a cylindrical reactor without air and saturated with MMA vapors.

Irradiation was done in the I.G.S.-3, a panoramic ^{60}Co irradiator at the Nuclear Engineering Department of the University of Palermo, having an activity of about 1400 Ci. The absorbed doses were 8, 15, 30 and 60 KGy. After irradiation, the samples were kept under dynamic vacuum at 40°C until not polymerized MMA was totally eliminated. As a rule, during the time between irradiation and tests, all samples were kept at a temperature of about -18°C to inhibit further radical reactions.

Pure PMMA was obtained irradiating MMA, in absence of air, at the same total absorbed doses. Also pure EVA-AA samples were irradiated, under vacuum, at the same total doses.

MMA loading and insoluble fractions determination

MMA loading after gamma-rays polymerisation was determined by the weight difference of rubber sheets before MMA swelling and after irradiation. In order to extract loaded MMA, irradiated samples were subjected to soxhlet extraction using acetone as solvent, typical solvent for pure PMMA and non-solvent for pure EVA-AA. Solubility tests were done in soxhelet using toluene, which is solvent for both EVA-AA and PMMA copolymers. The extraction time was always about 48 hours.

Mechanical tests

Mechanical tensile tests were done by an Instron machine mod 1122. Compression moulded specimens with a gauge length of 4 cm were strained at a initial strain rate of 0.05 s^{-1}. Young modulus (E) and elongation at break (ε_b) were reported as average values for at least five samples.

Transparency tests

Trasmittance measurements, in the visible region, was performed with a Perkin-Elmer Lambda 2 UV/VIS Spectrophotometer.

Dynamic mechanical analysis

Dynamic-mechanical thermal analysis was performed, for all prepared blends and for pure omopolymers, through a DMTA by Rheometrics. The frequency was 31.4 Hz and the temperature range was between -60°C and 150°C. The heating rate was 1.5°C/min. A rectangular geometry was used (width 13 mm, length 15-25 mm, thickness 1 mm) at a strain amplitude of 0.2%.

Scanning electron microscopy

Micrographs were obtained using a Jeol-JSM 5600 LU Scanning Electron Microscope. The samples, fractured under liquid nitrogen, were coated with gold under vacuum to make them electrically conductive.

Results and Discussion

In Table 1 the weight percentage of MMA loaded in the rubber after irradiation is reported. It can be observed that the total absorbed irradiation dose does not affect significantly the loaded amount.

Table 1. Weight percentage of MMA loaded in the rubber after irradiation

Dose (KGy)	MMA in EVA-AA (wt %)
8	39
15	40
30	41
60	40

All irradiated samples were subjected to soxhlet extraction using acetone, typical solvent for PMMA and non-solvent for EVA-AA; the results are reported in Table 2. The experimental results indicate that almost all MMA loaded in the rubber though irradiation remains trapped in the matrix and this amount does not depend on the irradiation dose.

Table 2. Weight percentage of MMA not extracted from the rubber after irradiation

Dose (KGy)	MMA not extracted (wt %)
8	96
15	97
30	94
60	98

In Table 3 solubility data for blends and pure polymers subjected to similar irradiation procedures are reported. For blends large amounts of insoluble fractions are detected, while pure PMMA and rubber are completely soluble. It is noticeable that the total insoluble fractions are greater than the corresponding fractions relative to MMA component, reported in table 1, thus indicating the formation of a strongly interacting structure which hinders the solubility of the rubber component.

Table 3. Solubility data for blends and pure polymers

Dose (KGy)	EVA-AA (wt %)	PMMA (wt %)	EVA-AA+PMMA (wt %)
8	Soluble	Soluble	38
15	"	"	49
30	"	"	43
60	"	"	48

Mechanical tensile results for blends are reported in Table 4, together with data relative to pure unirradiated polymers. With respect to unirradiated rubber, all blends present a significant increase of elastic modulus, while even though in presence of a decrease of the elongation at break, the material maintains its ductile behaviour. Consequently a generalised significant increase of tensile strength, with respect to pure unirradiated rubber is also achieved. For the effect of the irradiation dose, the trend is a generalised, not dramatic, decrease of elongation at break and an increase of modulus and tensile strength with the dose until 30 kGy. The decrease of modulus value for the material irradiated at 60 kGy can be attributed to the incoming occurrence of degradation phenomena.

Table 4. Mechanical tensile results for blends and pure polymers

SAMPLE	Young modulus (MPa)	Tensile strength (MPa)	Elongation at break (%)
EVA-AA unirradiated	18	9.5	875
EVA-AA+MMA 8 KGy	290	17.89	300
EVA-AA+MMA 15 KGy	332	19.43	275
EVA-AA+MMA 30 KGy	380	21	258
EVA-AA+MMA 60 KGy	243	19.63	217
PMMA unirradiated	1728	50.5	4

In fig. 1 the transmittance in the UV-visible wavelength range for blends and pure polymers is shown. It is worth to notice that the transparency of EVA-AA MMA blends is slightly lower than the corresponding values of pure PMMA, and significantly higher with respect to pure rubber.

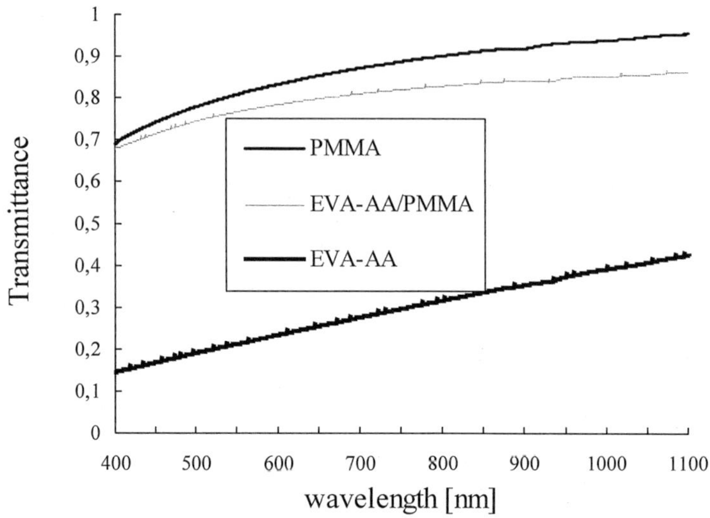

Fig. 1. Transmittance in the UV-visible wavelength range for the blend and pure polymers

Information about the blend structure have been obtained by dynamic mechanical analysis and scanning electron microscopy. In fig. 2 dissipation factor, tan δ, is reported as a function of temperature. Two relaxations can be clearly observed; indicating of the presence of two incompatible components. On the basis of the temperature values of both relaxation, we can hypotize that the polymer components are radiation modified EVA-AA rubber and polymethylmethacrylate formed by the irradiation of MMA monomer entrapped in the rubber. The presence of two distinct peaks is an indication of the absence of compatibilization phenomena; but if we compare the transition temperatures of both components with the pure corresponding polymers, see fig. 3, we can observe the presence of interactions that affect each other. In fact glass transition of both rubber and

PMMA components increases with respect to pure polymers, further indication of the presence of strong interactions between the two components.

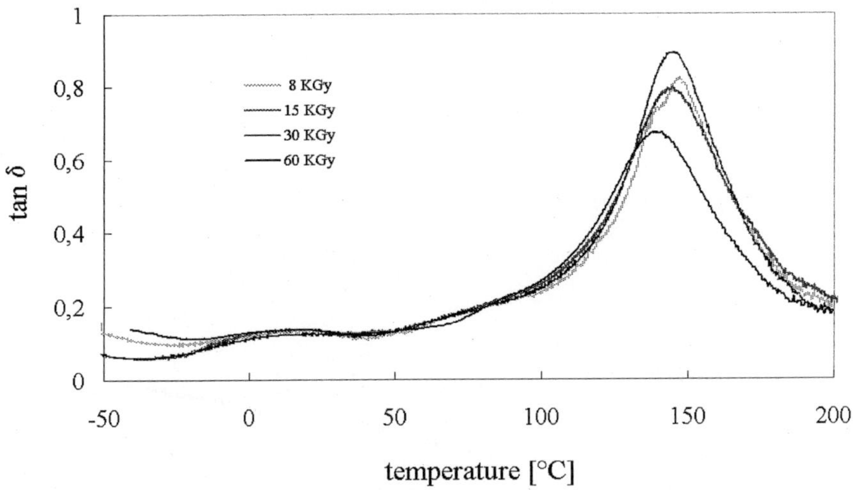

Fig. 2. Dissipation factor, tan δ, versus temperature.

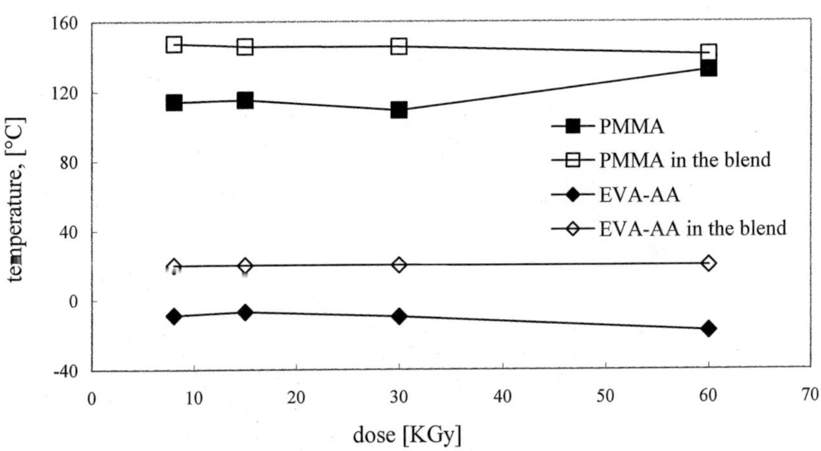

Fig. 3. Transition temperatures of both blend components and of the pure polymers versus the dose.

All these results are well in agreement with morphologycal analysis, shown by micrographs in fig. 4a-b, where scanning electron microscopy of just fractured and formic acid etched surfaces of EVA-AA/MMA blend is reported. Not etched surface does not allow to distinguish the presence of separated phases, while etching evidences the presence of very small voids indication of a very distributed PMMA phase inside the rubber matrix.

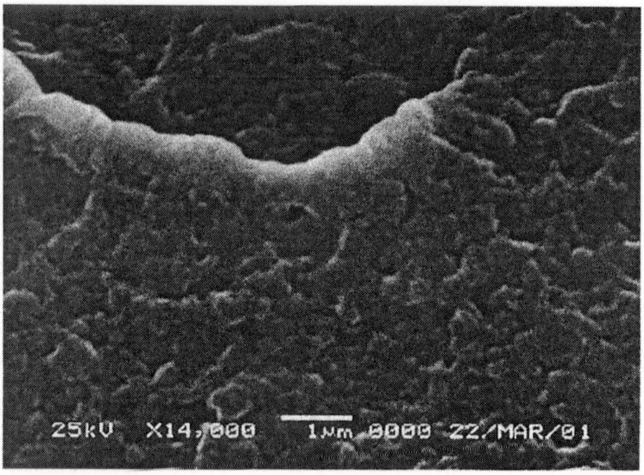

Fig. 4a. SEM micrographs of just fractured of EVA-AA/MMA blend.

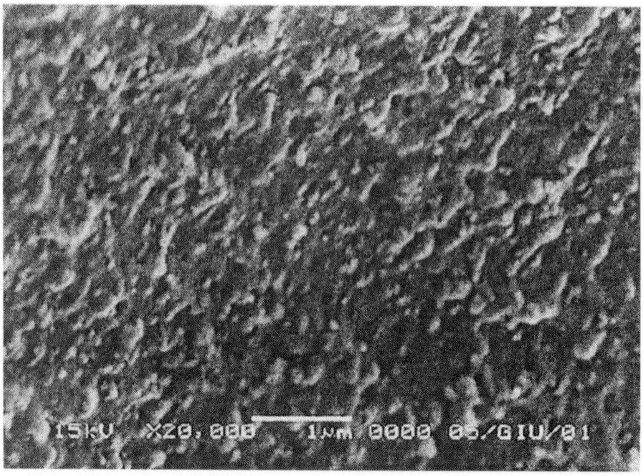

Fig. 4b. SEM micrographs of formic acid etched surfaces of EVA-AA/MMA blend.

Conclusions

In this work EVA-AA based blends are obtained through gamma radiation of MMA swollen rubber matrices. The aim is to obtain a strongly "compatibilised" system favouring the interaction of polymerising MMA with rubber.
Solubility tests indicate that the irradiated rubber-MMA swollen systems have significant insoluble fractions, despite the complete solubility of both pure polymers subjected to the same irradiation procedures. Mechanical tensile results are in between the properties of EVA-AA and PMMA polymers, with a good ductility even though in presence of significant increase of modulus value, with respect to rubber. Dynamic mechanical behaviour are typical of incompatible systems with two distinct relaxation for both components; however the temperature values indicate the presence of strong interactions between the two components, in line with solubility and mechanical behaviour.

Acknowledgement

This work has been financially supported by the CNR grant n 99.01205.CT03

[1] Z. Oommen, G. Groeninckx, S. Thomas, *Journal of Polymer Science Part B: Polym. Phy.* **2000**, 38, 525
[2] M. Akay, S.N. Rollins, *Polymer*, **1993**, 34, 1865
[3] V. Mishra, F.E. Du Prez, E.J. Goethals, L.H. Sperling, *J. Appl. Pol. Sci.* **1999**, 58, 347
[4] P. M. Remiro, C. C. Riccardi, M. A. Corcuera, I. Mondragon, *J. Appl. Pol. Sci.*, **1999**, 74, 772
[5] D. W. Clegg, A. A. Collyer *"Irradiation Effects on Polymer"* Elsevier, London 1991
[6] J. W. T. Spinks, R. J. Woods *"Introduction to radiation Chemistry"* John Wiley, New York 1990
[7] C. Cangialosi, C. Lindsay, P. T. McGrail, G. Spadaro, *European Polymer Journal*, **2001**, 37, 535
[8] D. Cangialosi, P. Fuochi, M. Lavalle, P.T. McGrail, G. Emmerson, G. Spadaro *Radiat. Phys. Chem*, in press.

Macromol. Symp. ***180****, 43–59 (2002)*

Influence of "Controlled Processing Conditions" on the Solidification of iPP, PET and PA6

Vincenzo La Carrubba (1), Valerio Brucato (2), Stefano Piccarolo (1)*

(1) D.I.C.P.M., University of Palermo, Viale delle Scienze, 90128, Palermo, Italy
(2) D.I.C.A., University of Salerno, Via Ponte don Melillo, 84084 Fisciano (SA), Italy

Summary: In this work reliable experimental data for three semicrystalline polymers (iPP, PA6, PET) crystallised under pressure and high cooling rates are supplied. These results were achieved on the basis of a model experiment where drastic "controlled" solidification conditions are applied. The final objective was to quantify the effect of two typical operating conditions (pressure and cooling rate) on the final properties and morphology of the obtained product. The influence of processing conditions on some macroscopically relevant properties, such as density and micro hardness is stressed, together with the influence of processing conditions on the product morphology, investigated by means of Wide Angle X-Ray Scattering (WAXS).
Results on the iPP samples display a decrease of density and micro hardness, due to the pressure increase, in a wide range of cooling rates (from 0.01 to 20 °C/s). PET samples exhibit an opposite behaviour with density and micro hardness increasing at higher pressures in the whole range of cooling rates investigated. PA6 samples behave similarly to PET displaying a less significant increase of density and micro hardness with pressure than PET samples.

1. Introduction

Solidification process in the most widely employed industrial processes is a complex phenomenon where flow fields, high thermal gradients and high pressures determine the final morphology and the resulting properties. Due to the experimental difficulties, any investigation of polymeric structure dependence upon pressure and cooling rate has been mainly performed using conventional techniques, such as dilatometry and differential scanning calorimetry under isothermal conditions or non isothermal conditions but at cooling rates several orders of magnitude lower than those experienced in industrial processes.

The complexity of the study is far higher if the behaviour of a semicrystalline polymer is investigated, since the lack of information on the influence of processing conditions on crystallization kinetics restricts the possibilities of modelling and simulating the

 CCC 1022-1360/00/$ 17.50+.50/0

industrial material transformation processes. The development of a model able to describe polymer behaviour under drastic solidification conditions is a goal very far to be achieved standing the actual background. Therefore a possible approach, along this general framework, consists in designing and setting-up model experiments with the aim to isolate and study the influence of some experimental variables on the final properties and on the final morphology. Thus a systematic investigation on polymer solidification under processing conditions should start on the separate study of the influence of flow, pressure and temperature on crystallization kinetics. In the past few years many researches have been devoted to determine how the final properties of a product depend upon cooling rate, that represents the most relevant parameter governing the polymer solidification. In the last few years a special equipment has been developed and widely tested in order to quench polymeric samples in a wide range of experimental conditions up to very high cooling rates (up to 2000 °C/s) under quiescent conditions: it has been possible to collect many information about the influence of cooling rate on the final properties of some widely used polymer, such as iPP, PET, PA6 [1-6].

The aim of the present investigation is to highlight the dependence of the final property distribution (in terms of mechanical properties, volumetric properties and final morphology) upon the experimental conditions adopted throughout the solidification. A model experiment able to achieve these goals under controlled conditions was designed, set-up and optimised in our laboratory. The methodological approach of recording the thermal histories of samples cooled under pressure and then analysing the resulting morphology and properties was adopted. All the available data concerning polymer crystallization under very high cooling rates at ambient pressure were extensively used as a starting point to evaluate the effect of pressure when superimposed to the one of cooling rate. Then non-isothermal crystallization behaviour was then determined as function of the applied pressure [7-9].

In order to generalise the results achievable with the help of this proposed experimental approach, three different polymers that covers a large variety of application fields has been adopted: isotactic PolyPropylene (iPP), PolyEthyleneTerephtalate (PET) and PolyAmmide6 (PA6). The study of three very different polymers, as far as their chemical and physical features are concerned, could be helpful for successive modelling of polymer behaviour during crystallization under processing conditions.

2. Experimental Part

For the purpose of our work an injection moulding iPP grade named "T30G" kindly supplied by Himont was employed, having the following main features: $M_n = 75100$; $M_w = 483000$; $M_w / M_n = 6.4$. PET was kindly supplied by SINCO. It is named "base PET" and its intrinsic viscosity is equal to 0.62 dl/g. PA6, kindly supplied by DSM, was characterized by M_w=25000, M_n =13000.

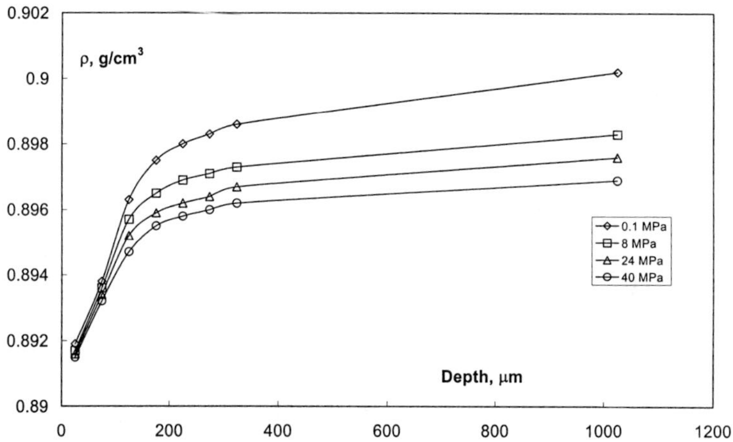

Fig. 1. Density depth profile for iPP (diaphragm 3.5 mm thick)

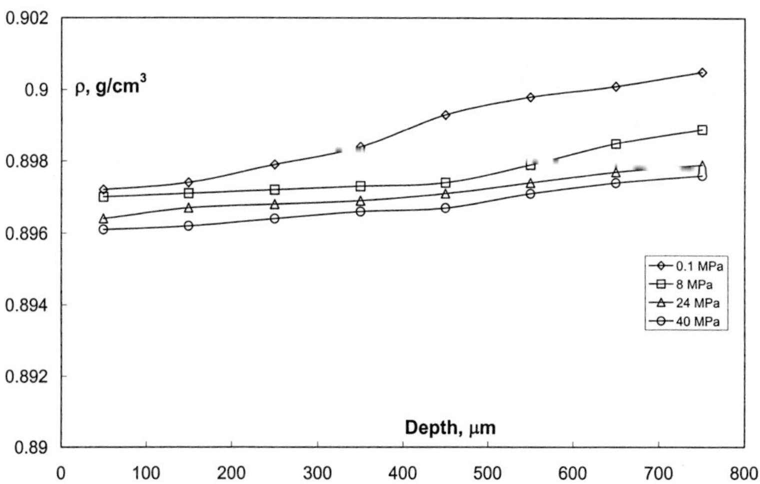

Fig. 2. Density depth profile for iPP (diaphragm 8 mm thick)

Materials were quenched under pressure and high cooling rates according to a procedure already reported [7-11]. The basic idea was to set up a model experiment where to couple rapid cooling histories with pressure levels up to 40 MPa under controlled conditions. The experimental apparatus consisted of a modified injection moulding machine, used as a source of molten polymer supplied at a pre determinable and maintainable constant pressure [7-8]. Another apparatus was used for crystallizing iPP samples at low cooling rates, consisting in a pressurized cell where the polymer was confined and maintained under hydrostatic pressure [9].

Prepared samples were subjected to the following characterization techniques: density measurements, Wide-angle X-ray Scattering (WAXS), Micro Hardness measurements. A deconvolution technique on iPP and PA6 WAXS patterns was applied to evaluate the final phase content of samples, according to a procedure reported elsewhere [12].

3. Results and Discussion

3.1 Density and micro hardness results on iPP samples

Results of density measurements on iPP samples are reported in figs. 1-2. Four different pressure conditions have been explored: 0.1 MPa, 8 MPa, 24 MPa, 40 MPa and two different diaphragms, 3.5 mm and 8 thick respectively. In fig. 1 the density depth profile for the 3.5 mm thick diaphragm is reported, whereas in fig. 2 the density depth profile for the 8 mm thick diaphragm is shown. Samples obtained with the 3.5 mm thick diaphragm are characterized by a surface cooling rate (at 70°C) of about 100°C/s; samples solidified using the 8 mm thick diaphragm have a surface cooling rate of about 20°C/s at 70°C. The cooling rates

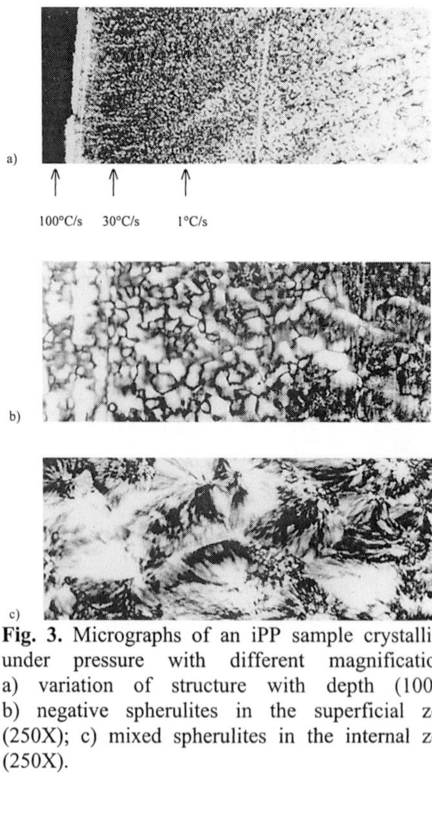

Fig. 3. Micrographs of an iPP sample crystallized under pressure with different magnifications: a) variation of structure with depth (100X); b) negative spherulites in the superficial zone (250X); c) mixed spherulites in the internal zone (250X).

have been evaluated at 70°C since it has been shown by several previous work in this field that the cooling rate calculated at 70°C is a good measure of quench effectiveness [1-2]. It is worth noticing that a diaphragm thickness corresponds to a superficial cooling rate since the diaphragm acts as a "thermal insulated body" [7-9].

Figs. 1 and 2 show that in both experiments, for all pressure values, density is increasing proceeding from the surface to the bulk of the sample. This behaviour can be related to an increase of crystallinity considered that the internal slices are cooled at lower rates. In other words, internal layers of the polymer undergo a less severe quenching and hence a larger crystallinity develops giving rise to a larger density. In figs. 1 and 2 the largest part of the density increase takes place in the neighborough of the surface and this is independent on the applied pressure. This is qualitatively confirmed by the micrographs shown in fig. 3 a) where the significant changes of the formed structure are localised near the surface in a layer of about 400 μm. It is also worth noting that we find negative spherulites near the cooled wall [fig. 3 b)], whereas inside the sample there is the typical spherulitical structure characterized by mixed spherulites. [fig. 3 c)]. Negative spherulites have a greater birefringence than mixed spherulites of the internal sample zone, depending upon the higher cooling rates experienced by the initial sample layers.

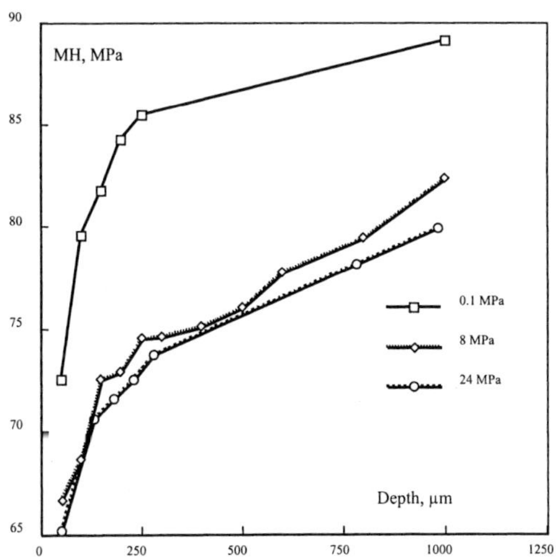

Fig. 4. Micro Hardness depth profile for iPP

Both figs. 1 and 2 show the same somewhat unexpected result: as a matter of fact density decreases when pressure increases at the same depth. The decrease in the density by effect of pressure is minimum at the sample surface and grows with depth. Furthermore, the majority of the density change is observed by varying the pressure from 0.1 to 8 MPa, this value being quite low especially if compared to the typical pressure values attained in polymer processing. This experimental result may be of relevance for modelling shrinkage and internal stresses distribution in injection moulding; particularly important is the fact that this outcome is more pronounced in the bulk of the sample [13-15].

In fig. 4 micro hardness measured along the depth on a transverse section versus depth is reported. The same dependence on pressure as density is observed, although the curves at pressures above 8 MPa collapse on each other, which is also relevant for process simulation especially with reference to mechanical properties.

As a matter of fact also from fig. 1 and 2 it is worth noticing that on increasing pressure density curves tend to level off. Additionally, the greatest density differences are in the internal zones of the samples, which are subjected to the lowest cooling rates. Going from the surface towards the bulk of the sample (i.e. decreasing the cooling rate) the effect of pressure on the final density tends to disappear. This is also relevant for process simulation.

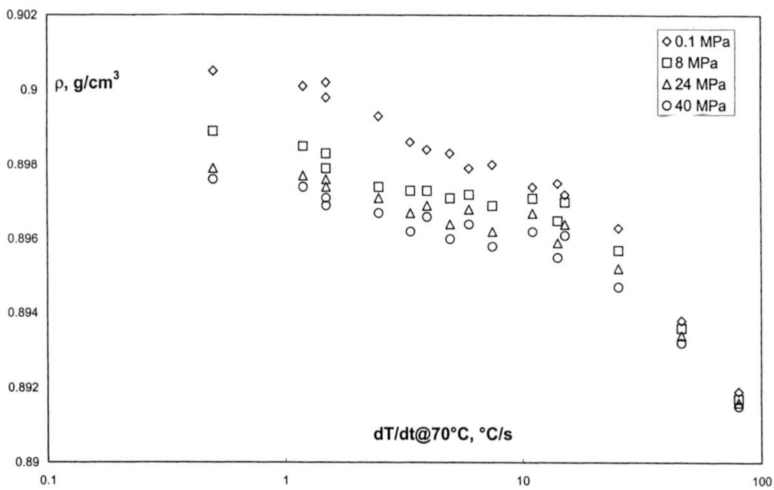

Fig. 5. Density versus calculated cooling rate for iPP

Fig. 5 is obtained by plotting the density data of figs. 1 and 2 versus the cooling rate calculated at 70°C by using a transient heat conduction model, as reported elsewhere [7-8]. The value of the calculated cooling rate was averaged across every slice thickness (50μm). The use of the transient model is also validated by the quite satisfactory overlapping of the data referred to a surface cooling rate of 100 and 20 °C/s. Also fig. 5 shows that at constant cooling rate final density decreases with pressure. Independently on solidification pressure the material shows the same trend with the cooling rate, with a density drop above 10-20 °C/s. Finally, fig. 5 shows that the decrease of density with pressure vanishes with increasing cooling rate. This implies that the influence of pressure is more pronounced in the bulk of the sample, which is highly relevant in process simulation.

Also micro hardness results of fig. 4 were converted into the diagram shown in fig. 6, where micro hardness as a function of cooling rate is reported. In this case the local value of the calculated cooling rate was adopted.

A similar pressure dependence on the density has also been observed by *He and Zoller* [16] using a standard dilatometer by *GNOMIX Inc.* to measure sample specific volume during crystallization from the melt. A constant slow cooling rate (2-3 °C/min) under constant pressure was applied bringing the sample back to a fixed pressure at the end of the test (10 MPa). It is worth noticing that most experimental results provide the specific volume at the measurement pressure, whereas in our case we measure density at ambient pressure after solidification under pressure. *He and Zoller* observed an increase of specific volume with increasing crystallization pressure in the case of iPP while PA66 and PET show an opposite behaviour [16]. Although they tried to explain the density decrease with the formation of γ phase, which is less dense than α phase, in our samples there is no evidence of γ phase formation [7-9]. Their results re however in agreement with our results, where final density (measured at atmospheric pressure) of samples solidified under pressure turns out to be lower than the one of samples solidified at atmospheric pressure.

3.2 Density versus cooling rate for iPP from 0.01 to 100 °C/s, extension to low cooling rates

In order to extend the range of cooling rate to be explored below the minimum achievable cooling rate using the modified injection moulding machine (around 1°C/s), the already mentioned pressurized cell was used, completely different from the former as for the way of producing a constant pressure field during sample crystallization. In both experiments (modified injection moulding machine and pressurized cell) the same methodological approach of recording the thermal history experienced by the samples during cooling and then analysing the resulting sample morphology was adopted. Due to the small size and the low achievable maximum cooling rate (less than 2°C/s), the samples crystallized in the pressurised cell were completely homogeneous and their density was measured in a density gradient column.

Reporting all the data obtained using both apparatus, the density versus cooling rate (evaluated at 70°C) dependence shown in fig. 6 is obtained. It is worth reminding that the cooling rate was measured for samples crystallized in the pressurized cell, whilst for samples solidified in the modified injection moulding machine the cooling rate was calculated by means of a heat transfer model. The first result that must be emphasised is the good matching of data referred to completely different experimental set-up (fig. 6). As a matter of fact, the pressurised cell data obtained by solidification at 2°C/s (maximum achievable cooling rate with this apparatus due to the intrinsic constraints imposed by the heat transfer), indicated with open symbols, exhibit a satisfactory overlapping with the data obtained in the modified injection moulding machine, indicated with full symbols. This outcome confirms once more the reliability of the whole experimental route adopted, including the use of the simplified heat-transfer model to map the distance from the diaphragm with cooling rate [8].

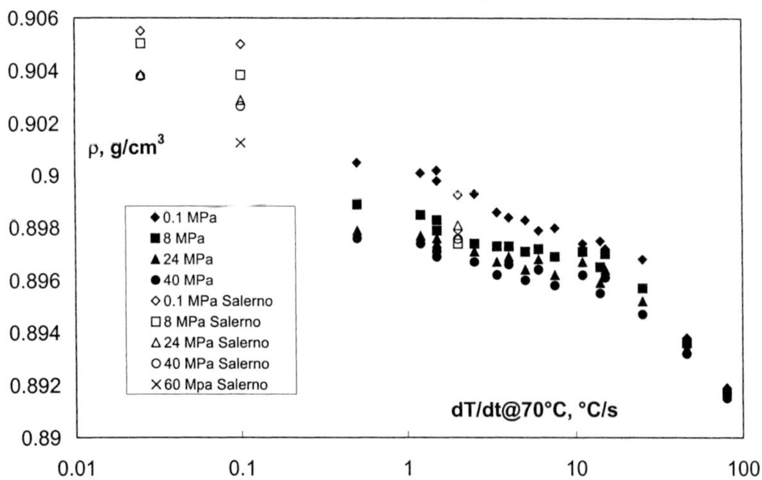

Fig. 6. Density versus cooling rate of samples prepared in the injection moulding machine (full symbols) and in the pressurised cell (open symbols)

The other experimental points achieved by means of the pressurised cell were at the cooling rates of 0.1 °C/s and 0.02°C/s (indicated with open symbols in fig. 6). The final density versus cooling rate dependence covers nearly four orders in magnitude in cooling rate (from about 0.01 up to nearly 100°C/s). Three regions can be distinguished within the diagram. A region of high cooling rates (higher than 20°C/s), characterised by a sharp fall in density, in this zone the role played by the solidification pressure is almost negligible, since the density gap between 0.1 and 40 MPa is very small. In other words, in the high cooling rate region, the final density value is dominated by cooling rate, and it can be therefore considered to be independent on the pressure applied during the crystallisation process. In the second region, spanning from 1 to 20°C/s, density exhibits a gradual fall at all the investigated solidification pressures. Solidification pressure here strongly affects the final density value, the higher the pressure the lower the final density.

In this range of cooling rates it may be observed that a cooling rate increase has the same effect as an increase of pressure. On this basis the effect of pressure on the final density may be accounted for by using a "master curve" which reports density as a function of a so-called "equivalent cooling rate", correctly taking into account the real cooling rate and the additional shift caused by the solidification pressure [7]. The physical basis of this approach is that, across this zone, an increase of cooling rate at a constant pressure corresponds to an increase of pressure at a constant cooling rate in terms of final density and phase distribution (from WAXS). The third region corresponds to the low cooling rate zone, ranging from 0.01 to 1°C/s. A gradual decrease of density may be noticed with the same influence of solidification pressure on final density seen in the region 1-20°C/s. The effect of pressure on the final density may be accounted for by using a different master curve, determined on the basis of the experimental density data of this region. From fig. 6 it is also evident a sharp fall of density from 0.1 °C/s to 1°C/s at all the solidification pressures. This result together with the WAXS deconvolution data lead to the conclusion that a further process must occur when cooling rate becomes lower than 1°C/s. This process is responsible for the observed increase of density and alpha phase.

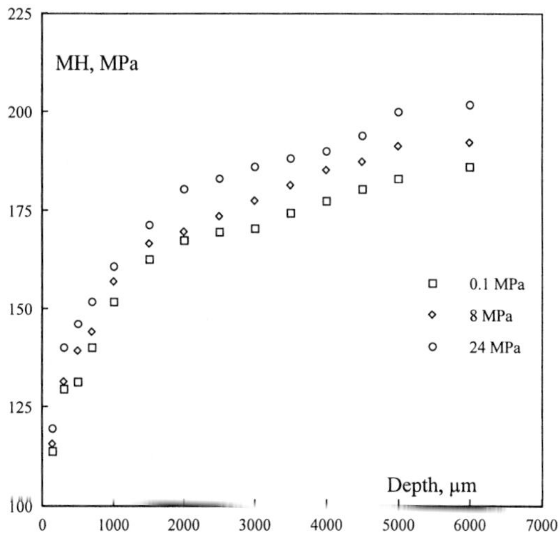

Fig. 7. Micro hardness versus depth (PET)

A "secondary crystallization" process in series to the first one may indeed give a plausible justification of density and WAXS results, as explained elsewhere [9].

3.3 Micro hardness and density measurements on PET samples

On PET samples it was not possible to measure the density variation by microtoming thin slices in the direction of sample thickness, due to the high brittleness of specimens which did non allow one to microtome them. Anyway it was possible measuring density of thick layer of polymer extracted from the bulk of the samples for all the adopted working pressures: 0.1, 8 and 24 MPa. Furthermore many density data of samples solidified under atmospheric pressure in a wide range of operating conditions are available. These measurements can lead to an indirect calculation of density data based on the densities of samples crystallized under ambient pressure conditions.

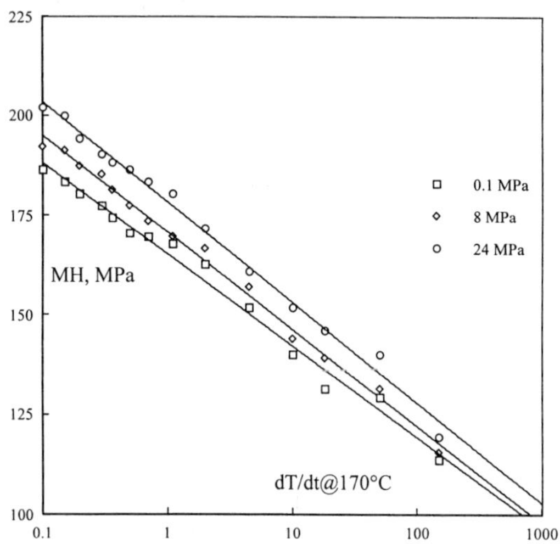

Fig. 8. Micro hardness versus cooling rate (PET)

On samples crystallized under pressure several micro hardness measurements were performed. A standard Vickers microindenter was used. A load of 100 g was applied for 6 s. with a loading velocity equal to 50 g/s. In this way a complete mapping of the sample was possible. Experimental results, in terms of micro hardness as a function of depth for all solidification pressures, are reported in fig. 7. First of all, as in the case of iPP, micro hardness increases going from the cooled surface to the bulk of the sample. In other words internal layers of polymer undergo a less severe quenching, which determines a lower value of crystallinity. This in turn causes micro hardness to increases on increasing depth, as shown in fig. 7.

As for dependence on pressure, PET displays an opposite behaviour with respect to the one exhibited by iPP (see fig. 4). As a matter of fact an increase of pressure corresponds to an increase of MH. Any possible explanation of the different behaviour displayed by the two polymers should probably rely on the different chemical structure. For PET *Phillips and Tseng* [17] have found similar results on PET crystallized under pressures up to 200 MPa. They observed a reduction of the average spherulite size as pressure is increased, with a much greater effect than that observed in other studies on different polymers. A crystallinity level nearly double than the one obtained at atmospheric pressure was obtained at a pressure of 150 MPa. It appears from their results that the increase of density with pressure is a direct result of the applied pressure during crystallization, since PET is a polymer with a very rigid repeating unit. Probably the molecular flexibility could be the key to understand the different behaviour displayed by iPP and PET under pressure. Furthermore the present results are in agreement with the experimental results presented by *He and Zoller* [16], who found a significant increase of density with pressure for PET, in contrast with the decrease observed for iPP.

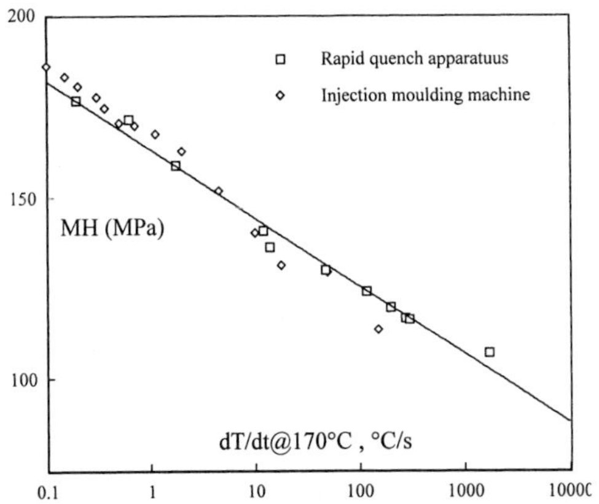

Fig. 9. Comparison between micro hardness data at 0.1 MPa obtained in different apparatus

From fig. 7, by using the "depth/cooling rate transformation function" for PET [9], calculated by using the heat transfer model, one can obtain data shown in fig. 8. In this figure micro hardness versus the calculated cooling

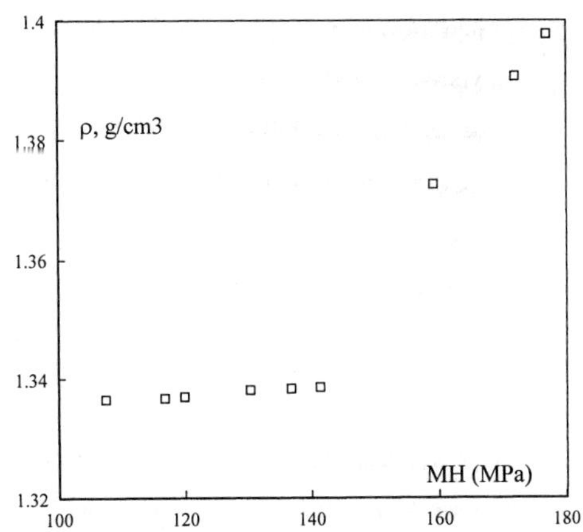

Fig. 10. Density versus micro hardness for samples obtained in the rapid atmospheric quench apparatus

rate (measured at 170°C) is plotted. Cooling rate is calculated at 170°C for PET since other works the past few years have demonstrated that this parameter is a good measure of quench effectiveness [5]. Fig. 8 shows that micro hardness decreases when the cooling rate increases. Furthermore, at a constant cooling rate, micro hardness increases with pressure in the whole cooling rate range examined. Finally micro hardness goes almost linearly with cooling rate (in a log scale) for all the solidification pressures.

It is interesting to compare micro hardness data of samples crystallized at ambient pressure in the apparatus developed for this work with micro hardness data of samples crystallized in a fast quenching ambient pressure apparatus developed by the authors few years ago. It is worth remembering that in the low pressure apparatus thin polymeric films (100 μm) are obtained and therefore the temperature profile inside the specimens is nearly flat. As a consequence of that, the cooling rate values are directly measured by using a thermocouple that records the thermal history. In our case the cooling rate is calculated by using the transient model. The comparison, shown in fig. 9, is quite satisfactory, demonstrating once again a good agreement between data obtained with the two different apparatus.

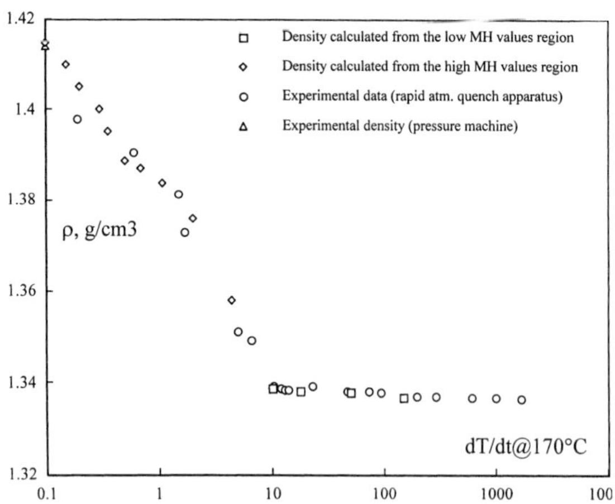

Fig. 11. PET. Comparison between experimental and calculated density (on the basis of micro hardness data)

Fig. 10 reports density versus MH for films obtained in the rapid quench apparatus at ambient pressure. In the low-density range, there is a significant increase of MH being the density nearly constant. Above micro hardness values of 140-150 MPa, the increase of micro hardness is accompanied by a parallel increase of density. This relationship between density and micro hardness can be described by using two different straight lines: ones in the low-density region and one (with an higher slope) in the high-density region. By using this analytical representation it is possible to calculate in an indirect way some density data. As a matter of fact micro hardness data reported in fig. 8 can be transformed into density data, shown in fig. 11. In fig. 11 are also reported experimental density measurements of samples solidified in the low pressure quench apparatus. Experimental and calculated data are closed each other, the comparison being satisfactory. In the same figure there is also one experimental point at a cooling rate of 0.1 °C/s. This point refers to a density measure made on a thick slice extracted from the bulk of the sample crystallized at 0.1 MPa in our apparatus. The obtained vale is also in agreement with the other results.

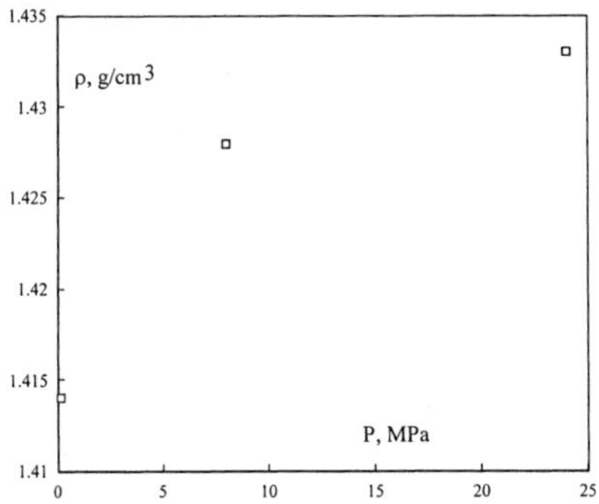

Fig. 12. PET density at 0.1 °C/s versus pressure

Finally fig. 12 shows density data for samples extracted from the bulk of samples crystallized under different pressure (having a cooling rate of 0.1 °C/s) as a function of solidification pressure. As a further proof of PET behaviour under pressure, density increases with pressure.

3.4 PA6 density and MH dependence upon cooling rate and pressure

Fig. 13 reports density of PA6 samples as a function of depth for all the adopted solidification pressures. Three different zones can be identified on that diagram. The first zone, near the cooled wall, is characterised by higher cooling rate values, and it could be easily noticed that density, going from the surface up to a depth of around 700 μm, remains nearly constant at a constant solidification pressure. Only a very small gap separates experimental densities of samples

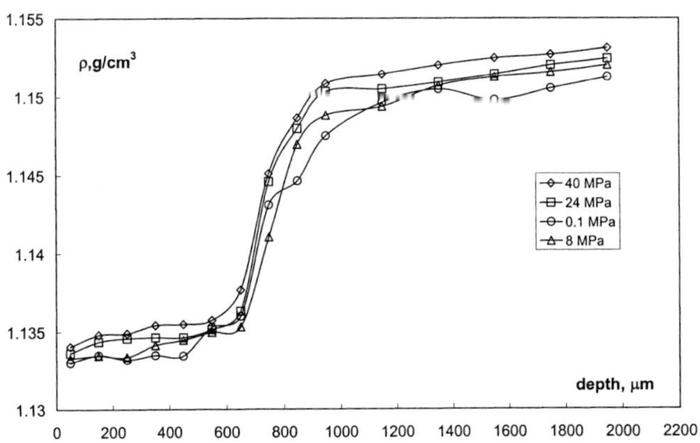

Fig. 13. PA6 Density depth profile

crystallised at ambient pressure with respect the ones crystallised under higher pressure conditions. Anyway the increase of density with increasing solidification pressure is not negligible. The second zone corresponds to a sort of "Transition zone" where the most significant variations of density versus depth take place. As a matter of fact, in the region where depth ranges from 700 μm up to 1000 μm, a sudden rise of density should be pointed out, independently on the working solidification pressure. In the whole transition zone the effect of pressure on the final sample densities turns out to be negligible. Finally, the third region, the "bulk zone", starting from a depth of approximately 1000 μm, it is the region where density tends to level off with increasing depth for a given solidification pressure. The effect of solidification pressure on the final density across this zone appears to be more pronounced than in the low-density plateau. Specifically, the higher the pressure the higher the resulting final density. On the whole, the increase of density due to the increase of solidification pressure is of the same order of magnitude of the decrease of density observed in the case of iPP, as already discussed above. This evidence turns out to be in full agreement with the dilatometric results reported by *He and Zoller* in 1997 [16], who found a decrease of density with solidification pressure of 0.1% in the case of iPP, while they noticed an increase of density of 0.3% in the case of PA66. It should be remembered that the cooling rate adopted during their experiments was very low (1-2°C/min) due to the intrinsic constraints imposed by the apparatus [16].

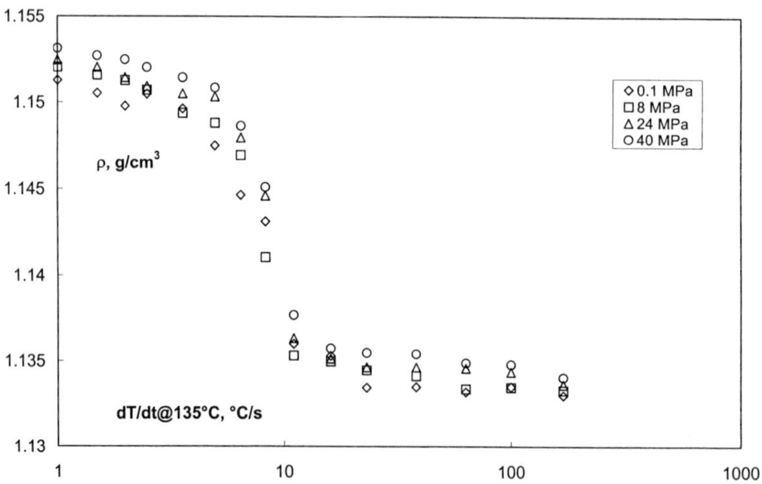

Fig. 14. PA6 density versus cooling rate evaluated at 135°C

If density of PA6 is reported as a function of cooling rate measured at 135°C [4] by means of the "depth-cooling rate mapping function" [9], the same trend discussed above is observed, as it could be noticed by looking through fig. 14. The low-density plateau corresponds to the high cooling rate region, and its starting point is located above 10°C/s. This region covers more than one order of magnitude in cooling rates, reaching the cooling rate of 250°C/s. The high-density plateau covers nearly one order of magnitude in cooling rate, going from the lowest achieved cooling rate during the experiment (1°C/s) up to 7-8 °C/s. As for the transition zone, it should be pointed out that it is located in a very narrow region, centred at a cooling rate of about 8-10°C/s. This cooling rate could be defined as a sort of "critical cooling rate" since small changes of cooling rate around this value could determine large structural and morphological variations that, in turn, determine large density variations. A very similar value of critical cooling rate was already found for a different PA6 some years ago, which is in agreement with the results here presented [4]. This value corresponds to the so-called "crystallizability" value defined by *Ziabicki* several years ago [18].

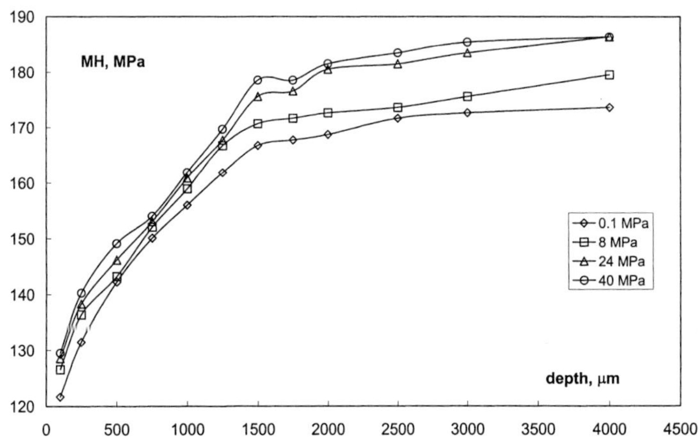

Fig. 15. PA6 micro hardness depth profile

As far as the micro hardness behaviour of PA6 is concerned, a different sensitivity with respect to the density behaviour is to be highlighted. Fig. 15 reports MH of PA6 samples as a function of depth for all the adopted solidification pressures. The first evidence to be analysed is a continuous gradual raise of micro hardness with depth, starting form the cooled wall. In other words, the low-density plateau, observed in the case of density, does not show up for micro hardness. A second issue concerns the high micro hardness plateau value. As a matter of fact this plateau corresponds only partially to the one observed in density, since its starting point is located at a much larger depth, around 1500 μm. All these differences could be explained by taking into account that density is a material "bulk" property, while micro hardness is a "surface" property and for this reason the latter turns out to be much more sensitive to any structural modification than the former. The reader should remember that a similar comparison was made in the case of iPP, where the structural and morphological modifications at the cooled wall can be easily highlighted by measuring the micro hardness profile (fig. 4), while in the case of density the variations measured near the surface were much less evident (figs. 1 and 2). As for the influence of pressure on the final micro hardness values, fig. 15 shows that an increase of pressure determines a parallel increase of micro hardness in the whole explored depth. Nevertheless the effect of pressure is more remarkable in the internal region of samples (in the high micro hardness plateau zone), where the gap between low pressure and high pressure crystallised samples tends to broaden. When Micro Hardness is reported as a function of cooling rate (fig. 16), the value of cooling rate above which the increase of micro hardness due to the increase of pressure turns out to be less pronounced, could be identified. This last corresponds to a cooling rate of about 5-6 °C/s lower than the critical value (8-10°C/s) found in the case of density.

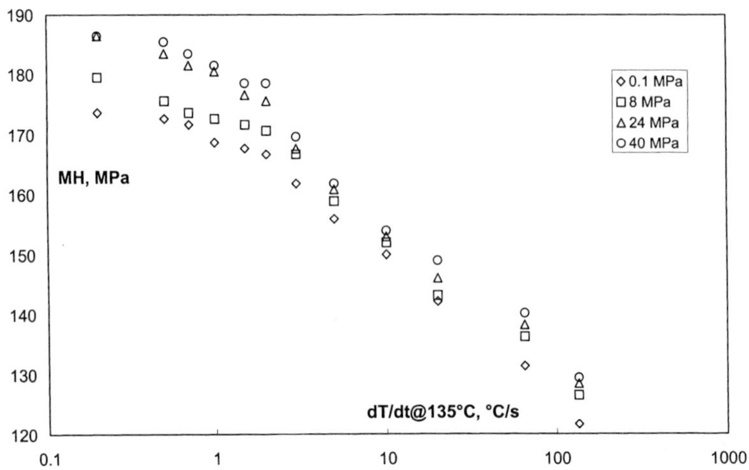

Fig. 16. PA6 micro hardness versus cooling rate evaluated at 135°C

References

[1] S. Piccarolo, *J. Macromol. Sci. – Phys. B*, **1992**, *31*, 501.

[2] S. Piccarolo, M. Saiu, V. Brucato, G.Titomanlio, *J. Appl. Poly. Sci.*, **1992**, *46*, 625.

[3] V. Brucato, T. Foresta, S. Piccarolo, *Proc. of NUPHIMAT'96*, **1996**, p.125.

[4] V. Brucato, S. Piccarolo, G. Titomanlio, *Int. Jou. of Forming Processes*, **1998**, *1*, 35.

[5] S. Piccarolo, V. Brucato, Z. Kiflie, *Polym. Eng. and Sci.,* **2000**, *40*, 1263.

[6] V. Brucato, G. Crippa, S. Piccarolo, and G. Titomanlio, *Polymer Eng. and Sci*, **1991**, 31.

[7] V. La Carrubba, V. Brucato, S. Piccarolo, *Polym. Eng. And Sci.*, **2000**, *40*, 11.

[8] V. Brucato, S. Piccarolo, V. La Carrubba, *Int. Polym. Proc.*, **2000** *15*, 103.

[9] V. La Carrubba, "*Polymer solidification under pressure and high cooling rate", Ph.D. Thesis*, CUES, Salerno 2000.

[10] G. Hitchcock, V. Brucato, S. Piccarolo, *Proc. of PPS12- Twelfth International Annual Meeting of the PPS*, **1996**, p.701.

[11] V. La Carrubba, *Thesis in Chemical Engineering*, University of Palermo, 1997.

[12] A. Martorana, S. Piccarolo, F. Scichilone, *Macromol Chem. Phys.*, **1997**, *198*, 597.

[13] G. Titomanlio, V. Speranza, V. Brucato, *Int. Polym. Proc.,* **1997**, *12*, 45.

[14] G. Titomanlio, S. Piccarolo, G. Levati, *Jou of Appl. Polym. Sci.,* **1988**, *35*, 1483.

[15] G. Titomanlio, A. Rallis, S. Piccarolo, *Polym. Eng. And Sci.*, **1988**, *29*, 209.

[16] J. He and P. Zoller, *J. Polym. Sci., Part B*, **1994**, *32*, 1049.

[17] P. J. Phillips, and H. T. Tseng, *Macromolecules*, **1989**, *22*, 1649.

[18] A. Ziabicki, "*Fundamentals of Fibre Formation"*, Wiley, London 1976.

Matrices for Mono-Material Composites (MMC). The Case of Poly(ethylene terephthalate)

Domenico Acierno [a], *Eugenio Amendola* [b], *Simona Concilio* [c], *Pio Iannelli* [c,*], *Antonio Massa* [d], *Paola Scarfato* [c]

[a] Dipartimento di Ingegneria dei Materiali e della Produzione, Università di Napoli, P.le Tecchio, I-80125 Napoli, Italy

[b] Istituto per la Tecnologia dei Materiali Compositi - CNR, P.le Tecchio, I-80125 Napoli, Italy

[c] Dipartimento di Ingegneria Chimica ed Alimentare, Università di Salerno, via Ponte Don Melillo, I-84084 Fisciano (Salerno), Italy

[d] Dipartimento di Chimica, Università di Salerno, via Allende, I-84081 Baronissi (Salerno), Italy

Summary: The use of the same thermoplastic polymer for both the matrix and the filler in the preparation of composites is very attractive, because such composites are in principle fully recyclable. Recently it has been reported the preparation of a mono-material composite (MMC) made of polypropylene to be used in automobile parts. In this work we report on the preparation and characterization of matrices for quasi MMC made of poly(ethylene terephthalate) (PET). Particularly we found PET copolymers containing isophthalate (IPA) units adequate to be used as matrix.

Keywords: poly(ethylene terephthalate); mono-material; composite; matrices; polycondensation

Introduction

In recent years a large interest has been focused on the synthesis and processing technology of new polymeric materials using a more environment-friendly chemistry. This corresponds to have recyclable materials with enhanced durability. The first "performance" is very important because the wide use of polymeric materials leads to a constant increase of their over-all production.

In the case of thermosets, which are used in applications requiring good mechanical performance and dimensional stability, recycling is not possible. The use of recyclable materials with similar performance may contribute in decreasing environmental impact.

Recently it has been reported the preparation of a mono-material composite (MMC) made of

 CCC 1022-1360/00/$ 17.50+.50/0

polypropylene [1] to be used in automobile parts. The use of the same thermoplastic polymer for both the matrix and the filler in the preparation of composites is very attractive, because such composites are in principle fully recyclable.

In this work we will investigate the possibility of producing an MMC of poly(ethylene terephthalate) (PET) and its copolymers. Particularly, we will focus our attention on the preparation and characterization of matrices for this kind of MMC.

Experimental Part

Matrices synthesis

All reagents were used as obtained from Aldrich. High crystalline PET (η_{inh} = 0.6 dl/g) fibers, cut 18 mm long, were obtained from Montefibre (Italy). Matrices were synthesized following three different procedures a, b, and c.

(a) oligomers. The general procedure for the synthesis of oligomers is the following: 3 g of Bis(hydroxyethyl)terephthalate (BHET) are heated to 230°C in the molten state and Ti(IV) isopropoxide is added [0.020%w]. Different molecular weight oligomers can be obtained by regulating reaction time. For Trimer synthesis a different procedure has been used, due to the difficulty in obtaining such a low molecular weight with the previous method: 12g of BHET with 0.90ml of pyridine are dissolved in 100cc of *o*-diclorobenzene at 140°C. 1.00g of terephthaloyl chloride is dissolved in 10cc of o-diclorobenzene. The latter solution is added drop wise to the former one. After two hours, 150cc of *n*-heptane is added and the powder is collected by filtration. The white powder is washed with *n*-heptane and with ethanol. The powder is dried in oven at 120°C [T_m=196.7 °C, ΔH_m=109.49 J/g].

(b) PET hydrolysis. BHET is added to PET at 240°C until the whole mixture melts to a homogenous fluid. Two drops of 30%$_w$ sulfuric acid are added and the reaction is carried out for 10 minutes. No vacuum is applied. The viscous fluid is recovered by pouring on an aluminum pad.

(c) Copolymers. 20 g of BHET with the required amount of isophtalic acid (IPA) are placed in a vial at 200°C and Ti(IV) isopropoxide is added [0.020%w], then the temperature is raised under vacuum. The whole reaction is performed as resumed, for some matrices, in Table 1.

Table 1. PET Copolymers (calorimetric data referred to the first heating run).

Synthesis	Thermal treatment			T_g/T_m [a]	ΔH_m [b]	M_w [c] ($x10^4$)	η^* x 10^3 [d]
	T (°C)	Time (min)	vacuum				
A(24) 16.7 $\%_w$ IPA	200 200 240 260 280	10 70 50 45 50	No Yes Yes Yes Yes	68/ -	-	63	0.66
A(25) 9.09 $\%_w$ IPA	200 240 280	60 60 60	No Yes Yes	74 / 232.2	37.5	35	/
A(26) 13.0 % IPA	200 240	60 180	No Yes	69 / 212.4	26.4		/
A(27) 16.7 $\%_w$ IPA	200 240	60 180	No Yes	67 / 197.7	7.2		0.034 for reology see Figures 9
A(28) 16.7 $\%_w$ IPA	200 240 280	60 60 120	No Yes Yes	68 / -	-		1.24 for reology see Figure 8
A(29) 16.7 $\%_w$ IPA	200 200 240 260 280	10 50 60 45 50	No Yes Yes Yes Yes	68 / -	-		
A(30) 28.6 $\%_w$ IPA	200 240	60 60	No Yes	68/-	-		
A(43) 9.09 $\%_w$ IPA	230 240 280	60 80 70	No Yes Yes	71/215	-		
A(44) 4.86$\%_w$ IPA	230 240 260 280	60 35 45 120	No Yes Yes	72/226	-		

[a] T_g,T_m /°C, glass transition and melting temperature, respectively; [b] ΔH_m/Jg^{-1}, melting enthalpy; [c] M_w / g mol^{-1}, average molecular weight; [d] η^* / Pa s, viscosity measured at 200°C with 5% strain at 10 rad/s.

Composites preparation

Matrix **A(30)**, previously dried at 120°C under vacuum for one night, is heated up to 200°C. Short crystalline fibers are added in the required amount and mechanically mixed with matrix for two minutes at 200°C. Then the mixture is pressed at 200°C for 30 seconds between two Teflon foils and quenched in a water bath. Composites using a PET fabric as filler were prepared in the same way with matrices **A(43)** and **A(44)**.

Characterization

Thermal measurements were carried out using a DSC-7 Perkin Elmer calorimeter under nitrogen flow at 10°C/min rate.

Polarized optical microscopy was performed using a Jenapol microscope.

Fiber diffraction spectra were recorded by means of a flat camera with a sample-to-film distance of 70.0 mm and the X-ray beam direction perpendicular to the fiber axis (Ni-filtered Cu-Kα radiation). The Fujifilm MS 2025 imaging plate and a Fuji Bio-imaging Analyzer System, mod. BAS-1800, were used for digitizing the diffraction patterns.

A Waters 150-C ALC/GPC instrument was used for GPC analysis, equipped with three columns PSS SDV from Polymer Standard Service (Gmbh) (PSS 100 Å- PSS 10^4 Å- PSS 10^5 Å).and a Jasco 875 UV detector operating at 254 nm . Polystyrene as standard and chloroform as solvent (samples previously dissolved in 1,1,1,3,3,3-hexa fluoro-2-propanol) at 1 ml/min and 30°C.

Dynamic mechanical tensile analysis (Single Cantilever) was carried out by means of a Perkin Elmer DMA 7 analyzer, at 5°C/min heating rate under nitrogen flow. Frequency was set at 1 Hz. Sample size is 18.0x8.4x0.3 mm.

The dynamic flow properties of the molten materials were measured with an ARES rheometer (Rheometrics, Inc.) using a parallel plates geometry (plate radius=12.5 mm, gap=1 mm). The samples for testing were dried at 105°C in a vacuum oven for 16 hrs and analyzed under a nitrogen gas purge in order to minimize thermo-oxidative degradation phenomena. Temperature sweep tests were carried out in the temperature range of 220÷180°C at a constant stress amplitude. Frequency sweep tests (0.1÷100 rad/s) were made at 200°C at constant strain amplitudes (5%, 10% and 20% strain).

Discussion

Preliminary investigation

In order to set matrix performances, filler (PET fibers) response to thermal treatment should be investigated first. Composite preparation will require the insertion of fibers in a molten matrix with a suitable viscosity to allow processing. Processing temperature should be as lower as possible to save tensile properties of fibers.

In Figures 1 and 2, DSC and DMA traces of a fiber sample of PET are shown, respectively. At temperature higher than the glass transition temperature (T_g), tensile modulus decreases reaching a plateau at 150°C. Finally, at temperature close to 250°C fibers collapse. Annealing at a temperature lower than 230°C-240°C does not significantly affect fiber orientation degree as confirmed by the X-ray diffraction pattern of fiber samples (see Figure 3). For practical purpose, we take 220°C as the upper limit of the over-all processing.

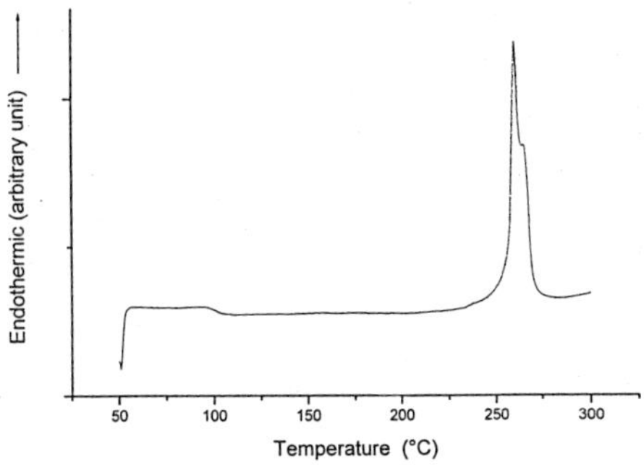

Figure 1. 1st heating run for a high oriented and high crystalline fiber sample of PET

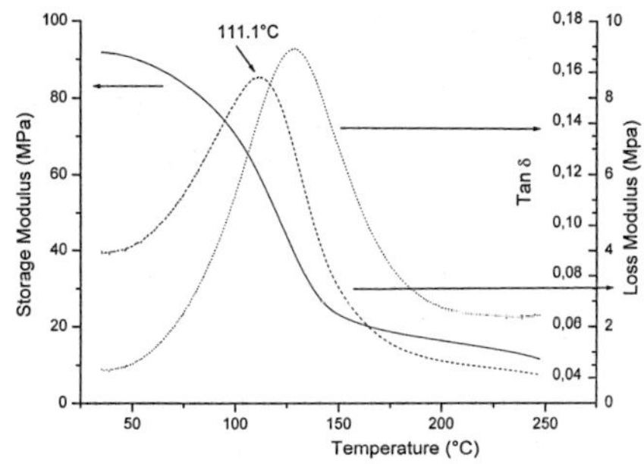

Figure 2. DMA traces for a high oriented and high crystalline fiber sample of PET (same sample of Figure 1) (Storage and Loss Modulus are given in arbitrary units)

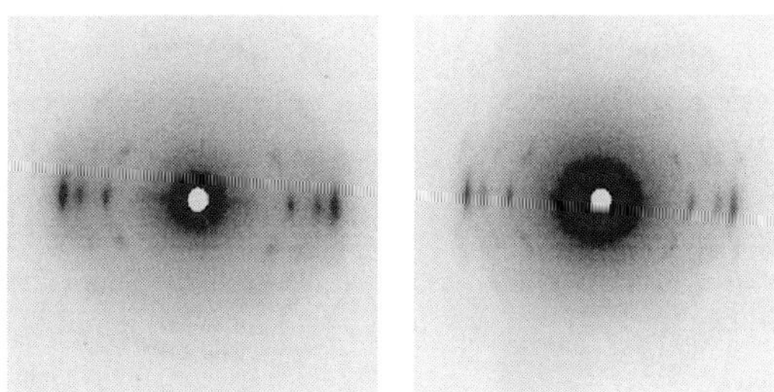

Figure 3. X-ray diffraction patterns of a fiber sample of PET (same sample of Figure 1) (right) as made and (left) after annealing at 220°C for 3 h

Matrix preparation

According to the experimental section, we synthesized matrices following three different approaches:

(a) Oligomers

In principle, fibers may be covered by a solid layer of oligomers taking advantage of the lower melting temperature (T_m) compared to polymer. Subsequent annealing at the appropriate temperature should increase molecular weight of matrix by solid state polycondensation, Ti(IV) as catalyst [2-6]. To avoid matrices desegregation due to glycol diffusion, oligomers with quite high molecular weight, but with low T_m, should be used. Unfortunately, this is not the case of PET oligomers. In fact, melting temperature increases suddenly with molecular weight as shown in Figure 4 (Figure 4 shows only a qualitative trend, low molecular weight compounds being out of the linearity range of GPC). For example, trimer has T_m=197°C and a mixture of oligomers with an average weight molecular weight (M_w) of 1.12 x 10^3 g/mol (see GPC in Figure 5) has T_m=248°C. Thus only very short oligomers (a mixture from trimer to pentamer) can be used as matrix but in this case the amount of glycol produced during polycondensation is too high. Consequently, trails with low melting oligomers ensure good covering of fibers but matrix desegregates during annealing at 180°C-200°C.

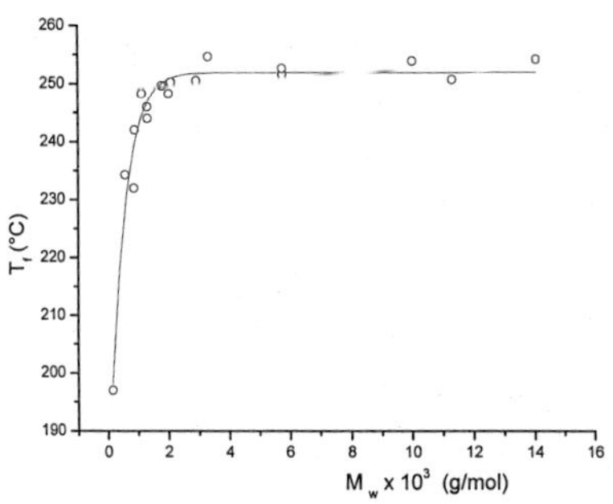

Figure 4. Melting temperature *vs.* molecular weight M_w (temperature referred to the second heating run)

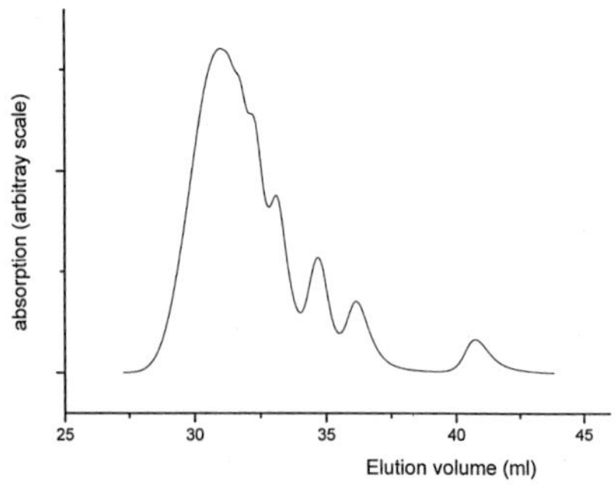

Figure 5. GPC curve for a PET oligomer with T_m = 248.3°C and M_w = 1.12 x 10^3 g mol^{-1}

(b) PET hydrolysis

Another way to decrease melting temperature is to introduce disorder along the molecular chain by inserting comonomers. This approach will be discussed in this and in the next paragraph, following two different procedures. Strictly speaking this does not correspond to have a pure PET matrix for preparing a real MMC, but the approach can be acceptable provided that composition differs as less as possible.

Here comonomers will be generated starting directly from PET by hydrolysis. In fact, hydrolysis, catalyzed by diluite sulfuric acid, is expected to cause: *i*) chain scission, generating carboxyl units; *ii*) vinyl units formation by dehydration of glycol; and *iii*) diethylenglycol (DEG) formation by etherification of two glycol terminal groups [7]. The first two cases produce a lower molecular weight polymer whilst the last one leads to a copolymer: in this case melting temperature depression depends on the amount of DEG. Unfortunately this corresponds to have also a decrease of glass transition temperature that has a negative effect on the tensile performance of matrix. The relative trend of T_m and T_g should be tested for different reaction conditions.

In order to decrease the temperature at which hydrolysis is carried out we prepare a mixture of PET/BHET by dissolving PET in BHET in the molten state. In this way, melting temperature decreases and hydrolysis can be performed at temperature ranging from 220°C to 240 °C.

Melting temperature *vs*. composition for hydrolyzed samples are reported in Figure 6 and Table 2. In spite of the different nature of the two components (high molecular weight PET and low molecular weight monomer BHET) the GPC curve consists of a single peak (see Figure 7). Then 10 minutes hydrolysis is enough to ensure, in terms of molecular weight, an homogenous matrix. Unfortunately , according to Figure 6 and Table 2, melting temperature is still too high. Increasing reaction time would decrease much more molecular weight thus acid hydrolysis seems to be not a suitable way for preparing matrices.

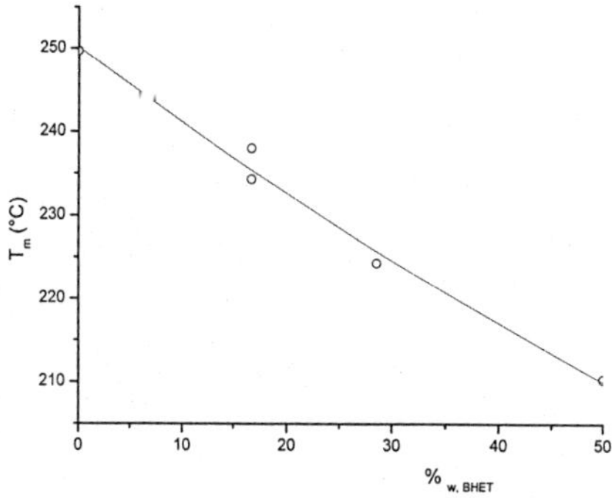

Figure 6. Melting temperature *vs.* composition of PET/BHET mixtures (temperature referred to the second heating run)

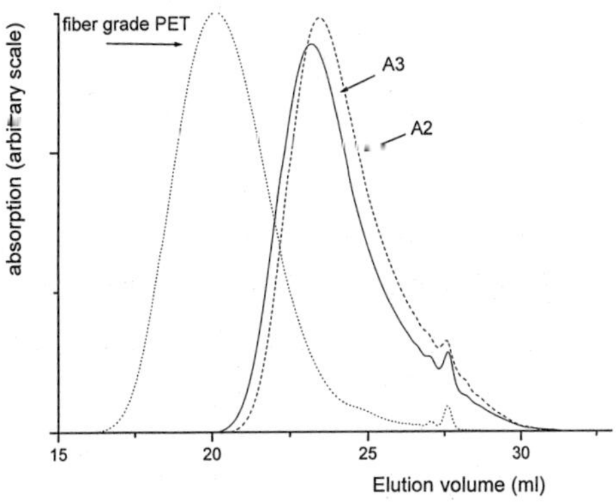

Figure 7. GPC elution curves of (a) **A(2)** [M_w = 6.8 x 10^3 g mol^{-1}], (b) **A(3)** matrix [M_w = 8.8 x 10^3 g mol^{-1}], and (c) fiber grade PET [M_w = 77 x 10^3 g mol^{-1}]

Table 2. PET hydrolysis (calorimetric data referred to the second heating run).

Sample	$\%_{w, BHET}$	T_m [a]	T_c [b]	ΔH_m [c]	M_w [d] ($x10^3$)
PET (pellets)	0	249.7°C	169.3°C	28.3	53.4
A(2)	50.0	210.3	158.7	26.2	6.8
A(3)	28.6	224.3	170.7	37.2	8.8
A(4)	16.7	238	190.0	39.1	
A(6)	16.7	234.3	184.0	39.9	

[a] T_m /°C, melting temperature; [b] T_c /°C, crystallisation temperature; [c] ΔH_m/Jg^{-1}, melting enthalpy; [d] M_w/g mol^{-1}, average molecular weight

(c) Copolymers

The last procedure to be tested is the synthesis of copolymers made of terephthaloyl and isophthaloyl units. It is well known that the insertion of conformationally rigid comonomers in the chain does not affect too much the glass transition temperature of PET. Moreover bend shaped monomer placed along the chain inhibits crystallization making lateral chain-to-chain register less favorite. This correspond to have a lower melting temperature. Isophathalic acid is one of these comonomers. The problem is to find out a matrix with good performances and a composition as close as possible to pure PET sample. Of course the insertion of more than two rigid bend comonomers might improve performances of matrix, even decreasing the total comonomers content.

In principle, there are several parameters which influence final matrix performances: composition, reaction temperature, reaction time, pressure, catalyst content, and so on. To reduce the number of variables we report on copolymers made of two monomers only, BHET and IPA, and some of reaction parameters fixed according to Table 1.

After several runs, we observed that:

i) at 200°C, IPA is not completely soluble in the molten BHET and a white suspension is obtained. After 60 min. suspension is still present; it disappears slowly only at higher temperature (230°C). If vacuum is applied, clarification of fluid is accompanied by sublimation of part of IPA, so that matrix composition is not well controlled;

ii) in the preliminary treatment at 200°C vacuum is not really necessary. Indeed atmospheric pressure is to be preferred because it avoids IPA and low molecular monomers sublimation;

iii) thermal treatment at 280°C brings to yellow and highly viscous matrix, thus lower temperature and longer reaction time should be imposed. The obtained matrices show viscoelastic behavior (see Figure 8) with the two G' and G'' curves quite superimposed. The obtained matrices show a non-Newtonian flow behavior (see

Figure 8): the complex viscosity curve is highly shear thinning in the whole frequency range examined and the two G' and G'' curves are quite superimposed in the lowest frequency region (ω = 0.1-5 rad/s);

iv) 60 minutes at 240°C were found adequate for having transparent colorless matrices with low viscosity (see Figure 9). Matrices with adequate viscosity might be prepared by regulating reaction time;

v) glass temperature transition is quite unaffected by IPA content (see dynamic tensile moduli in Figure 10).

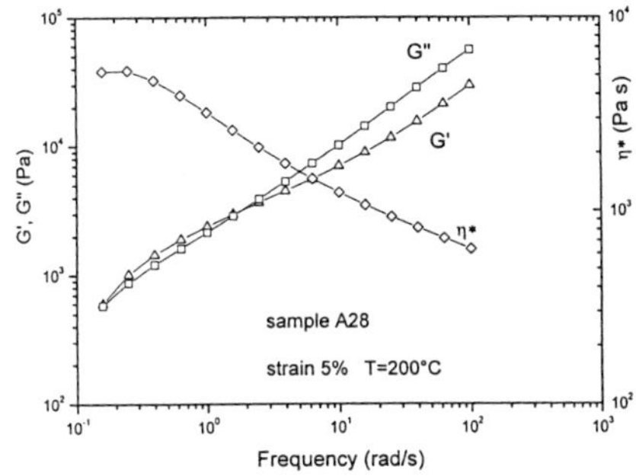

Figure 8. Dynamic flow properties of the molten **A(28)** sample (strain 5%)

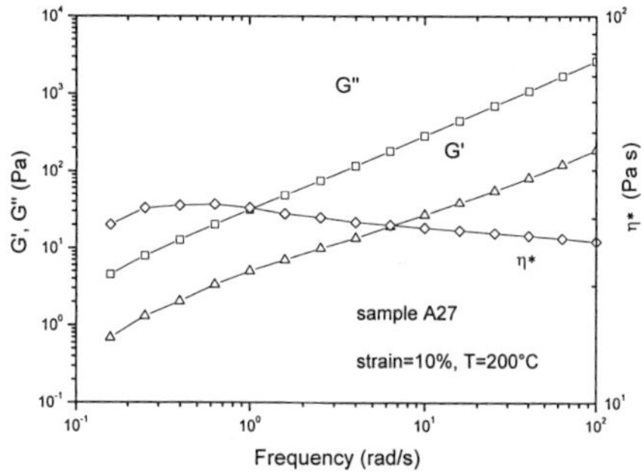

Figure 9. Dynamic flow properties of the molten **A(27)** sample (strain 10%)

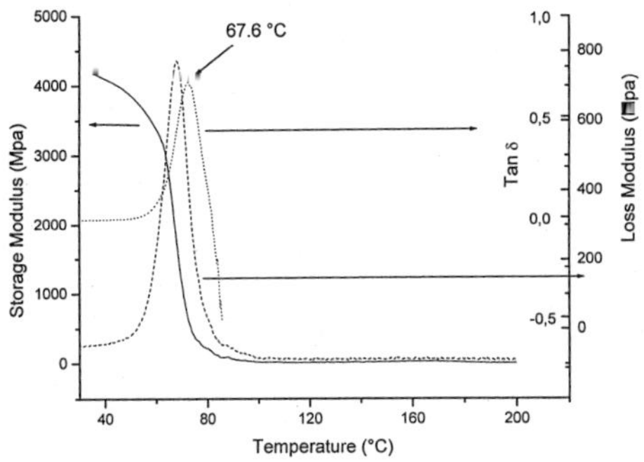

Figure 10. Dynamic tensile measurements for **A(24)** sample

Composites

Matrices **A(27)** and **A(30)** in Table 1 have been used for preparing PET composites at 200°C. In the case of **A(27)**, test was unsuccessful due to the partial crystallization of matrix occurring during thermal treatment at 200°C. Thus it was necessary to increase temperature up to 240°C, but at this temperature dissolution of fibers in the matrix occurs. In the case of **A(30)**, test was successful. In fact at 200°C **A(30)** is fluid and fibers can be easily added. Mixtures with different compositions were prepared and pressed at 200°C. At the present stage, we were not able to prepare homogenous composites, which are necessary for a correct analysis of tensile and mechanical performances, but we obtained only very preliminary samples. They were tested for analyzing fibers morphology: by X-ray diffraction analysis, fibers result to be still oriented and locally aligned (see Figure 11) and in the calorimetric traces the endothermic melting peak of fibers is visible (see Figure 12).

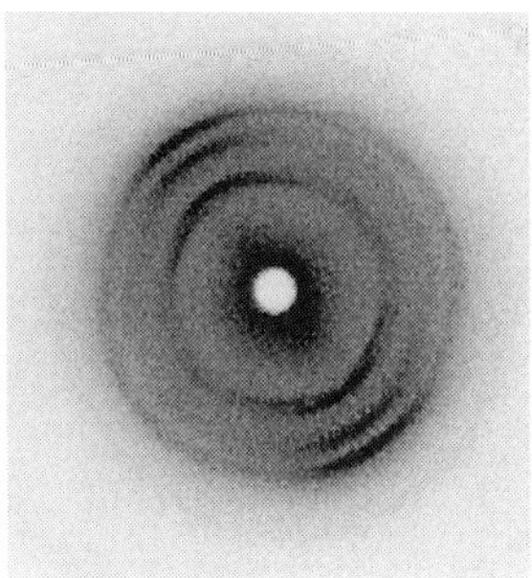

Figure 11. X-ray diffraction patterns of composite made of PET fibers as filler and **A(30)** sample as matrix

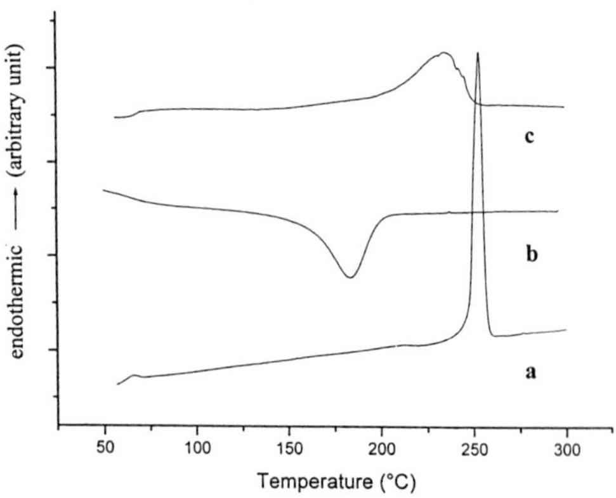

Figure 12. Calorimetric traces of a composite sample made of PET fibers as filler and **A(30)** sample as matrix: (a) first heating run, (b) first cooling run, and (c) second heating run

Conclusion

BHET/IPA copolymers are suitable to be used as matrices for PET composites.

We prepared a composite using matrix **A(30)**, but at the present stage of investigation we are not able to control homogeneity of final sample. Further work is necessary in order to prepare quite large samples with homogeneous distribution of fibers as filler. The amount of IPA should be reduced as much as possible by testing the synergic effect of different comonomers or crosslinking agents in order to have a quasi pure MMC. Moreover the effects of reaction temperature in the range 240°C-280°C and reaction time on matrices performances need a deeper investigation.

[1] E. Johannboeke and G. Popp, *Kunststoffe*, 89, (1999), p.77.
[2] J. Otton et al., *J. Polym. Sci.*, 26, (1988), p. 2199.
[3] T.H. Shah, J.I. Bhathy, G.A. Gamlen and D. Dollimore, *Polymer*, 25, (1984), p. 1333.
[4] B. Huang and J.J. Walsh, *Polymer*, 39, (1998), p. 6991.
[5] F.K. Mallon and W.H. Ray, *J. Appl. Polym. Sci.*, 69, (1988), p.1775.
[6] I. Devotta and R.A. Mashelkar, *Chem. Eng. Sci.*, 48, (1993), p. 1859.
[7] N. Halacheva and P. Novakov, *Polymer*, 36, (1995), p. 867.

Network Development During Epoxy Curing: Experimental Ultrasonic Data and Theoretical Predictions

*C. Pindinelli, G. Montagna[§], V.A.M. Luprano[§] and A. Maffezzoli**

Department of Innovation Engineering, University of Lecce
Via Monteroni 73100, Lecce, Italy; Email: alfonso.maffezzoli@unile.it
[§]PASTIS-CNRSM, Brindisi Italy

Summary: Ultrasonic waves propagation acts as a dynamic mechanical deformation on a material. When, during ultrasonic wave propagation, chemical or physical changes occurs, the evolution of an elastic modulus can be monitored. Therefore, this technique can be considered a powerful tool for non-destructive cure monitoring with a potential for "in situ" applications. In this work, the isothermal cure of a model epoxy resin cured with an amine is studied using propagation of longitudinal ultrasonic wave. The epoxy to amine ratio is optimized in order to reach full conversion of the amine groups during curing. The relative changes in the ultrasonic velocity and attenuation, measured by the transmission technique, have been applied to the calculation of the longitudinal modulus. The ultrasonic modulus has been compared with the degree of reaction measured using Differential Scanning Calorimetry (DSC). Furthermore a correlation between the ultrasonic modulus and the crosslinking density is presented combining DSC data with the stoichiometry of reactants according with the statistical theory of Miller and Macosko. The plot of the ultrasonic modulus as a function of the crosslinking density suggested that the theory of rubber elasticity can not be applied to the ultrasonic bulk longitudinal modulus as a consequence of the small deformation involved in the propagation of the ultrasonic waves.

Introduction

Ultrasonic inspection for non destructive analysis is routinely applied to the quality control presenting among other advantages the reliability and the cost. However, when chemical or physical changes in a material are associated with significant changes of the elastic constants, it is possible to characterize the evolution of their mechanical properties by means of acoustic measurements. Ultrasonic waves act as a high frequency dynamic-mechanical deformation applied to the material, leading to the determination of the loss and storage moduli and of the loss factor of the material. Only recently the potential application of acoustic wave propagation in the cure monitoring

 CCC 1022-1360/00/$ 17.50+.50/0

of thermosetting resins has been presented[1-4]. However, the correlation of the ultrasonic moduli changes with the development of the crosslinked structure has not been yet analyzed.

In this work, the isothermal crosslinking of a model epoxy resin cured with an exa-functional amine is studied using propagation of longitudinal ultrasonic wave. The epoxy to amine ratio is optimized in order to reach full conversion of the amine groups during curing. The relative changes in the ultrasonic velocity and attenuation, measured by the transmission technique, have been applied to the calculation of the longitudinal modulus. The ultrasonic modulus has been compared with the degree of reaction measured using a Differential Scanning Calorimeter (DSC). Furthermore a correlation between the ultrasonic modulus and the crosslinking density is presented combining DSC data with the stoichiometry of reactants according with the statistical theory of Miller and Macosko[5, 6].

Experimental

The studied resin system was a mixture of a DGEBA (diglycidyl ether of bisphenol A) based epoxy resin (Epon 828 from Shell) characterized by an epoxy equivalent weight (EEW) of 188 g/mol and an aliphatic exa-amine curing agent (Three Ethylene Tetra Amine) from Aldrich. The choice of an amine to epoxy ratio equal to 7.2 Phr (part of amine for a hundred part of resin by weight) is described below.

The cure is analysed using a differential scanning calorimeter (DSC) Perkin-Elmer DSC-7 operating in isothermal and non-isothermal conditions in presence of a nitrogen flux of 20 ml/min.

The longitudinal velocity and the ultrasonic wave attenuation during the cure at different temperatures are measured modifying the parallel plate rheometer Rheometric ARES. ARES oven and gap control system are used during ultrasonic measurements. Two custom made ultrasonic transducers, operating up to 100 °C, are fitted into the parallel plate disposable tools of the rheometer. The transducers are connected to a pulser-receiver system Panametrics 9100 providing a digital output and measurements are performed in the transmission mode filtering frequencies in the range of 2-4 MHz. The gap between the two transducers is measured using the ARES data acquisition software while ultrasonic data are processed in a separate computer and longitudinal velocity and attenuation are recorded every 9 s. A sketch of the apparatus is reported in Fig.1. The ultrasonic measurements were performed in isothermal conditions by placing the resin sample, in form of a thin sheet (less than 1 mm), into the preheated oven in contact with the transducers. Aluminum foils and a coupling gel are used for preserving transducer integrity and at the same time to optimize the ultrasonic coupling. The contraction of the resin during cure is measured and followed by the vertical movement of the upper transducer, driven by the gap control function of the rheometer. Isothermal ultrasonic and DSC measurements were obtained at the same temperatures.

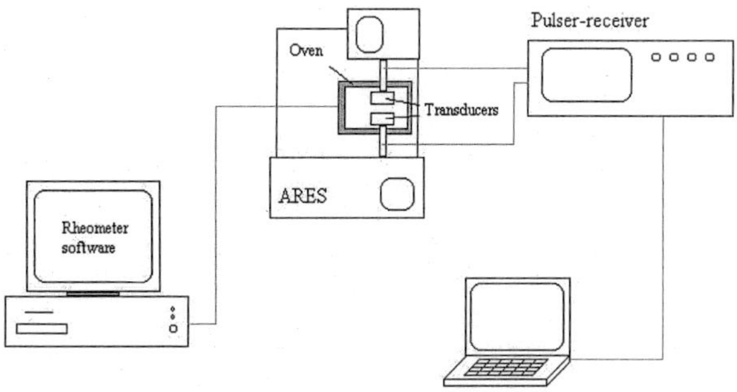

Figure 1. Sketch of the experimental apparatus for ultrasonic cure monitoring.

DSC analysis

Diffusion effects during cure

The degree of reaction (α_{DSC}) is calculated in isothermal DSC experiments following a classical procedure previously reported[4]. As shown in Table 1, the heat developed during isothermal DSC experiments (Q_{is}) is lower than maximum heat of reaction measured in non-isothermal scans Q_{tot}=428 J/g indicating that unreacted groups are still present.

Table 1. Heat of reaction, maximum degree of reaction and for isothermal curing of 7.2 Phr composition sample at various cure temperatures.

Cure Temperature	50°C	55°C	60°C	65°C	70°C
Q_{is} (J/g)	368.08	389.48	402.32	415.16	423.72
α_{max}	0.86	0.91	0.94	0.97	0.99
T_g	62°C	69°C	73°C	80°C	86°C

Curing rate of epoxy-amine reaction may be controlled either by chemical reactivity of functional groups either by diffusion of reacting species. During curing the reaction is initially controlled by chemical reactivity of amine and epoxy groups and the rate of reaction is independent of the diffusion constants of the species carrying these functional groups. As high crosslinking densities are reached, the diffusion of reactants significantly affects the reaction kinetics. In the last stage of the cure, epoxy-amine reaction is no long associated with small molecules migration, but it involves displacements of long branched molecules and segmental mobility of network. Therefore, steric hindrances and slow segmental diffusion lower the curing rate and lead to a reaction diffusion-controlled. Furthermore, segmental mobility is dramatically decreased by vitrification. In fact, the glass transition temperature of the growing network, continuously increasing during cure, approaches the isothermal cure temperature strongly reducing the molecular mobility. In these conditions, the reaction is slower and the cure, in practice, is arrested.

Isothermal data for degree of reaction (α_{DSC}) are obtained for curing of 7.2 Phr composition samples, lead at temperatures of 50°C, 55°C, 60°C, 65°C and 70°C. These data are analysed in order to show the dependence of curing reaction on diffusion as vitrification of resin is approached. As long as curing is controlled by chemical reactivity of reacting functional groups, the degree of reaction is described by the following differential equation:

$$\frac{d\alpha}{dt} = k(T)f(\alpha) \qquad (1)$$

where f(α) is a function of the degree of reaction and k(T) is the reaction rate constant. The dependence of k on absolute temperature T is governed by Arrhenius expression:

$$k = A\exp\left(-\frac{E_a}{RT}\right) \qquad (2)$$

where E_a is the activation energy and R is the gas constant (R=8.314 J/molK).

Assuming that the reaction is controlled by chemical reactivity, Eq.1 is valid at any temperature. A change of the experimental temperature affects the value of the rate constant, while the degree of reaction still follows Eq.1. Therefore, it is possible to superimpose the curves of degree of reaction as a function of time, obtained at different cure temperatures simply modifying the experimental time scale[7]. This time-temperature superposition is obtained using a shift factor, a_T, for the time scale assuming as reference temperature 70°C (a_T=1 for T=70°C). Table 2 shows the values of a_T for the different temperatures considered. The time dependence of the degree of reaction obtained in isothermal experiments at different temperatures is shown in Fig.2a while the superposition of curves resulting from the use of the ~~by the~~ shift factor a_T is reported in Fig.2b.

Table 2. Shift factor, maximum degree of reaction and degree of reaction at the transition to diffusion control at different cure temperatures.

Cure temperature	a_T	α_{max}	α_{lim}
50°C	0,26	0,86	0,75
55°C	0,38	0,91	0,79
60°C	0,50	0,94	0,82
65°C	0,72	0,97	0,85
70°C	1	0,99	-

As shown in Fig.2b, isothermal curves superimpose in the first stage of the process, when reaction is controlled by chemical reactivity; for higher values of α_{DSC} the reaction is slowered by the growth of the network and the reduction of segmental mobility due to vitrification. For high degrees of conversion the reaction becomes diffusion-controlled and superimposition is no long observed. This effect is stronger at lower reaction temperature, because diffusion is more difficult and the resin approaches vitrification temperature earlier as cure temperature is lower. As a consequence, the maximum value of the degree of reaction reached is higher for curing lead at higher cure temperature.

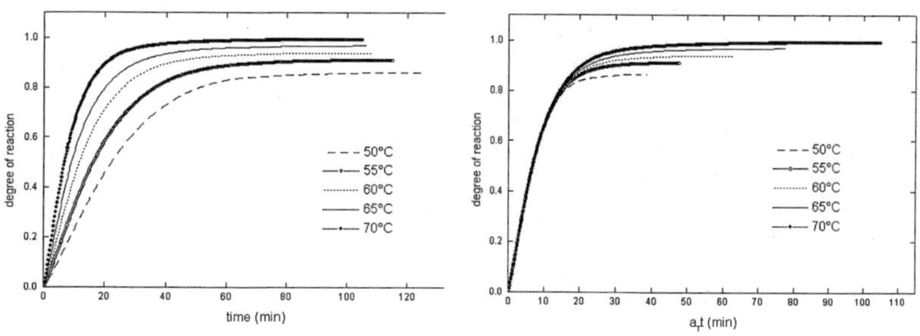

Figure 2. Plot of degree of reaction vs time: a) raw DSC data; b) superimposition of curves by shift factor a_T.

Table 2 reports the value of α at which reaction becomes diffusion controlled (α_{lim}) for each temperature considered. As shown in Table 2, although α_{lim} and α_{DSC} have been calculated following different procedures, they are characterized by the same increment with the cure temperature. This may be explained taking into account that the values of both α_{lim} and α_{DSC} result from the effect of diffusion on the reaction. Fig.3 shows the region of Fig.2b where the transition to the diffusion control mechanism occurs. Referring to the first stage of the reaction, when chemical reactivity control occurs, from eq.1 it can be obtained[7]:

$$\frac{k(T)}{a_T(T)} = \frac{\frac{d\alpha}{dt}}{f(\alpha)} = C_1 \tag{3}$$

where C_1 is a constant. Thus, at any temperature, the ratio of the rate constant and the shift factor is a constant and it does not depend on the temperature. Substituting Arrhenius equation for k(T), the following expression for a_T is obtained:

$$a_T(T) = \frac{1}{C_1}\exp\left(-\frac{E_a}{RT}\right) \tag{4}$$

Experimental data of shift factor obtained from isothermal measurements fit very well with Eq.4 and their analysis gives $C_1=1.295*10^{-8}$ and $E_a=12.413$ kcal/mol.

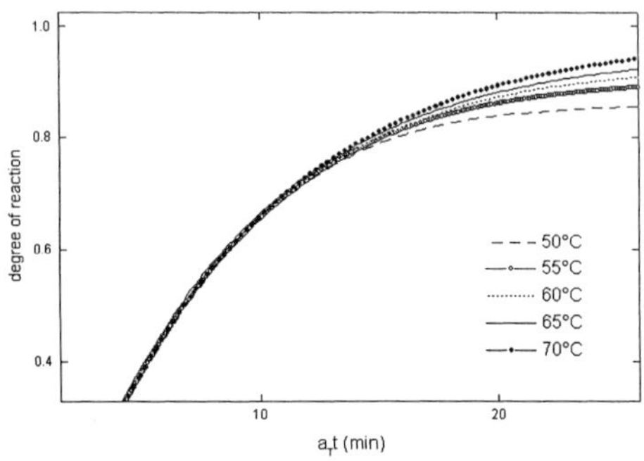

Figure 3. Transition from chemical reactivity control to diffusion control.

Therefore the development of a high density network may lead to incomplete reaction of amine or epoxy group as a consequence of the so called caging effect[8]. This effect is highlighted in Fig.4, where the heat of reaction for unit mass of amine during dynamic scans (Q_A) is reported as a function weight fraction of amine. As shown in this figure, for amine content lower than 8 Phr, a plateau is approached indicating full conversion of amine groups.

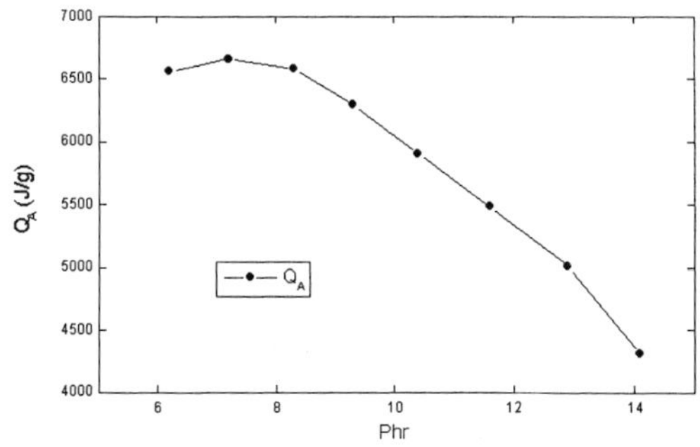

Figure 4. Heat of reaction for unit mass of amine (Q_A) as a function weight fraction of amine in Parts per hundred parts of resin (Phr).

Choice of amine/epoxy ratio

The theoretical stoichiometric ratio of amine to epoxy is 12.9 Phr. This results is confirmed by DSC analysis as the value corresponding to the maximum heat of reaction in a dynamic scan at 10°C/min. However, in view of a correlation of postgel network properties with ultrasonic data, the resin formulation must be properly chosen accounting for:

1. the maximum temperature for ultrasonic measurements (100 °C)
2. the value of the degree of reaction at the gel point (α_{gel}) depending on the functionality of amine (f_A) and epoxy (f_E) and from the stoichiometric ratio r[9]:

$$\alpha_{gel} = \frac{1}{\sqrt{r(f_A - 1)(f_E - 1)}} \tag{5}$$

3. the maximum value of the glass transition temperature (T_g) of the resin. T_g must be kept under 100 °C in order to perform at least an ultrasonic experiment leading to a final degree of reaction equal to 1.
4. Finally, as shown formerly in Fig.4, a full conversion of amine groups may be reached only for a large excess of epoxy.

On the basis of these considerations a formulation containing 7.2 Phr is chosen. For this formulation the T_g = 91 °C and α_{gel}=0.6 from Eq.5 and α_{DSC} may be assumed equal to the conversion of amine groups (p_A). This assumption represents a key point for the calculation of the crosslinking density through the statistical approach reported below.

Ultrasonic analysis

Velocity at five different temperatures are reported in Fig.5. The initial values decrease with temperature as expected[4] while the final values increase with the cure temperature. Since at higher temperatures are associated higher maximum values of the degree of reaction and hence higher crosslinking density, this result indicates that for the studied resin is more significant the effect of crosslinking compared with that of temperature.

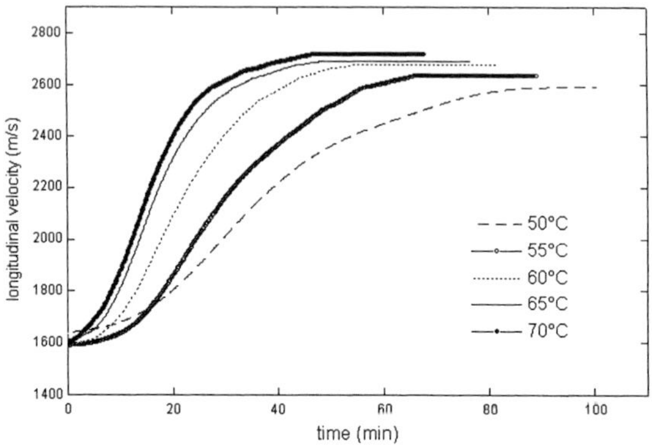

Figure 5. Longitudinal velocity measured at different cure temperatures.

The attenuation (Δ) is calculated measuring the amplitude of the ultrasonic waves in presence and absence of the sample and monitoring the thickness changes (d) during cure, according with:

$$\Delta = -\frac{1}{d(t)} \ln\left(\frac{A_{II}}{A_I} \frac{1}{T_{21}T_{12}} \right) \tag{6}$$

where A_I and A_{II} are the amplitudes measured without and with the sample between the transducers and T_{21} and T_{12} the transmission coefficients at the interfaces resin-transducer and transducer-resin respectively calculated from the acoustic impedances of the delay line of the transducer and of the resin[4].

The attenuation calculated using Eq.6 is reported in Fig.6. After a peak, corresponding to the vitrification as is detected at so high frequencies[3,4], the reduction of mechanical coupling between transducer and resin lead to a limited reduction of the attenuation detected in pulse echo measurements[4]. For this reason correct attenuation values are measured up to a small time interval after the peak.

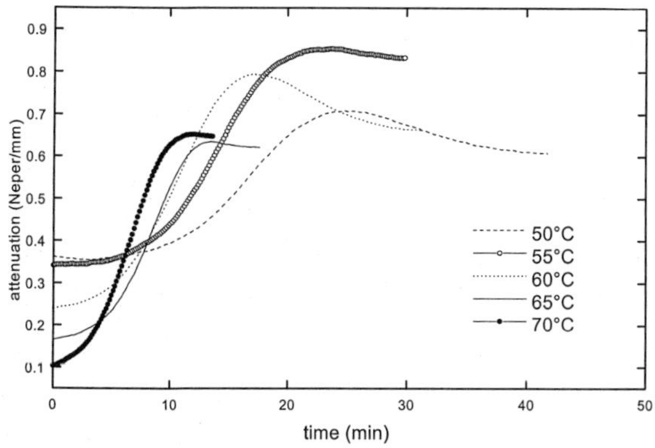

Figure 6. Ultrasonic attenuation measured at different cure temperatures.

When the sample dimension normal to the acoustic wave is large compared with respect to the wavelength, the wave propagation is governed by the complex bulk longitudinal modulus L* = L'+ iL". This complex modulus is related to the bulk (K*) and shear (G*) complex moduli[10] L*=K*+4/3G*. L' corresponds to the stiffness of a system that is deformed changing its dimensions in one direction while in the other two directions the dimensions are constrained to be constant, as it occurs in samples where two dimensions are much larger than the third. For a liquid or a soft rubbery material K'>>G' and L' may be considered equal to the bulk modulus, while K" and G" may be of the same order of magnitude and the effect of G" on L" cannot be neglected.

The measurements of ultrasonic longitudinal velocity (V_L) and attenuation (Δ) may be used for the calculation of the storage (L') and loss (L") bulk longitudinal moduli[9]:

$$L' = \frac{\rho V_L^2 \left[1 - \left(\frac{\Delta\lambda}{2\pi} \right)^2 \right]}{\left[1 + \left(\frac{\Delta\lambda}{2\pi} \right)^2 \right]^2} \qquad L'' = \frac{2\rho V_L^2 \left(\frac{\Delta\lambda}{2\pi} \right)}{\left[1 + \left(\frac{\Delta\lambda}{2\pi} \right)^2 \right]^2} \tag{7}$$

where ρ is the material density and λ is the wavelength, given by $\lambda = V_L/\nu$ as a function of the frequency, ν. For the studied case the effect of attenuation on the values of L' was negligible and a single longitudinal modulus $L = \rho V_L^2$ can be calculated. The difference between L and L', up to the attenuation peak, was lower than 3% and hence only L has been calculated.

A comparison between the degree of reaction calculated by DSC and L is reported in Fig.7 for one of the isothermal tests. At the beginning of the cure the degree of reaction obtained by DSC grows more rapidly than the longitudinal velocity as a consequence of the high concentration of reactive groups in the liquid reactive mixture. On the other hand, the growth of molecular weight in form of branched oligomers, has a limited effect on the mechanical properties of the resin before the gel point. Close to gelation, the growing branched molecules are characterized by extensive entanglements capable to provide an elastic response measured as a significant growth of longitudinal velocity and L. In the last part of the cure, a moderate or negligible increase of the degree of reaction is accompanied by a significant increase of L.

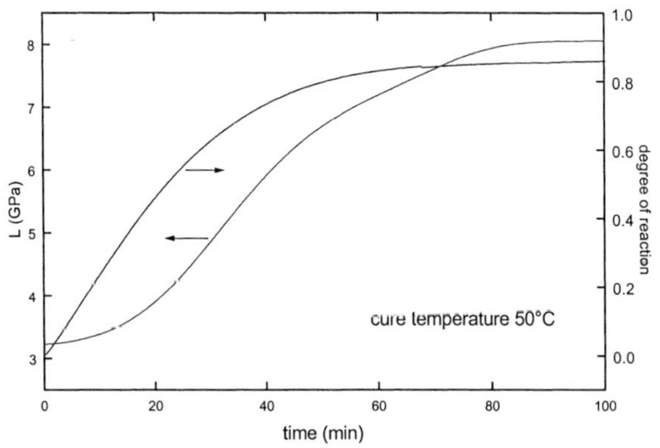

Figure 7. Comparison of time dependence of longitudinal modulus and degree of reaction.

Taking the values of L and α at corresponding times, the plot of L as a function the degree of reaction shown in Fig.8 is obtained for three cure temperatures. These data indicate that L is mainly sensitive to the increasing crosslinking density occurring after gelation rather than to the number of reacted groups. Therefore a strong non linear correlation between L and α is observed.

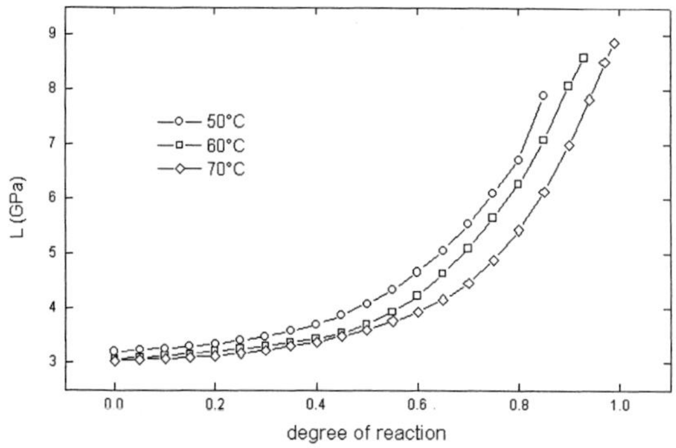

Figure 8. Longitudinal modulus as a function of the degree of reaction obtained combining DSC and ultrasonic data plotted in the former figure.

Nevertheless, the evolution of mechanical properties after gelation is theoretically predicted by Flory [9] in the theory of rubber elasticity and may be computed, for example, applying the statistical approach of Macosko and Miller[5,6]. Then L could be better correlated with the degree of crosslinking, defined as the concentration of effective junction points in the infinite network, rather than with the degree of reaction. The theory of Miller and Macosko for stepwise polymerization leads to the calculation of the crosslinking density (μ) and the concentration of active network chains (ν) through the determination of $P(F_A^{out})$, the probability that the event F_A^{out} occurs. For the studied system F_A^{out} is defined as the event that taking any amine group and moving out from the molecule to which the group is connected a dangling chain is reached. The theory derives an expression which provides $P(F_A^{out})$ as a function of the amine conversion p_A that, as mentioned above, can be assumed equal to α_{DSC}. Then for the studied formulation, characterized by r=0.556, the following equation is obtained:

$$rp_A^2 P(F_A^{out})^5 - P(F_A^{out}) - rp_A^2 + 1 = 0 \tag{8}$$

$P(F_A^{out})$ is calculated for different values of p_A using a numerical solution of Eq.8. Afterwards, an expression of $P(F_A^{out})$ as a fuction of the degree of conversion of amine groups is obtained by polinomial fitting. In particular for the postgel range of p_A (p_A>0.6) a simple third order polinomial expression is obtained:

$$P(F_A^{out}) = a_0 + a_1 p_A + a_2 p_A^2 + a_3 p_A^3 \tag{9}$$

where: α_0= 2.675, α_1= - 4.337, α_2= 3.262, α_3= -1.145.

A plot of $P(F_A^{out})$ vs conversion of amine groups as obtained from Eq.9 is shown in Fig.9. This expression is valid in postgel region only, where the infinite network is developed. For $\alpha<\alpha_{gel}$, in presence of molecules of finite size, $P(F_A^{out})$ is assumed equal to unity, as Eq.8 has no physical meaning.

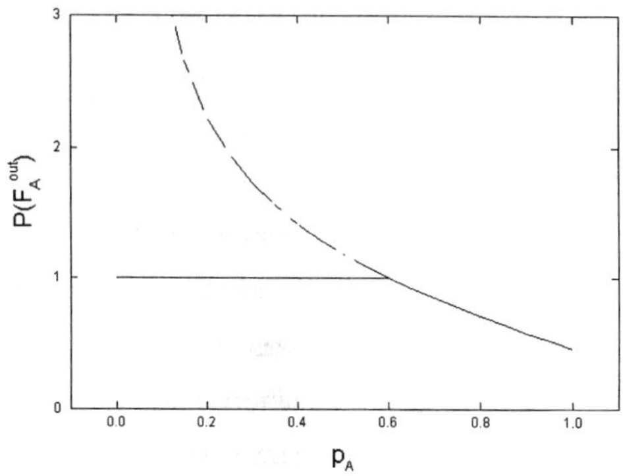

Figure 9. Plot of $P(F_A^{out})$ vs conversion of amine groups as obtained from Eq.9.

Once $P(F_A^{out})$ is calculated, the probability that an amine molecule is connected to m segments, belonging to the infinite network, is:

$$P(X_{m,6}) = \binom{6}{m} P(F_A^{out})^{6-m} \left[1 - P(F_A^{out})\right]^m \tag{10}$$

An amine molecule is an effective crosslinking point only if m>3. The value of $P(X_{m,6})$ are reported in Fig.10 for m ranging between 3 and 6; these values of m corresponds to the development of an effective crosslink. As a consequence of the large excess of epoxy, the probability of finding 6 reacted hydrogen of the amine monomer is very low and limited to the last part of the reaction.

Then the crosslinking density, μ, is given by:

$$\mu = [A_0] \sum_{m=3}^{6} P(X_{m,6}) \tag{11}$$

where $[A_0]$ represents the initial concentration of amine equal to: $5.52*10^{-4}$ moli/cm^3. The sum is between m=3 and m=6 being these values capable to generate a junction between 3 to 6 chains segments. Observing that at each junction point can be associated m/2 connected chain segments the concentration of the active network chains, ν, is obtained as :

$$\nu = [A_0]\sum_{m=3}^{6}\frac{m}{2}P(X_{m,6}) \tag{12}$$

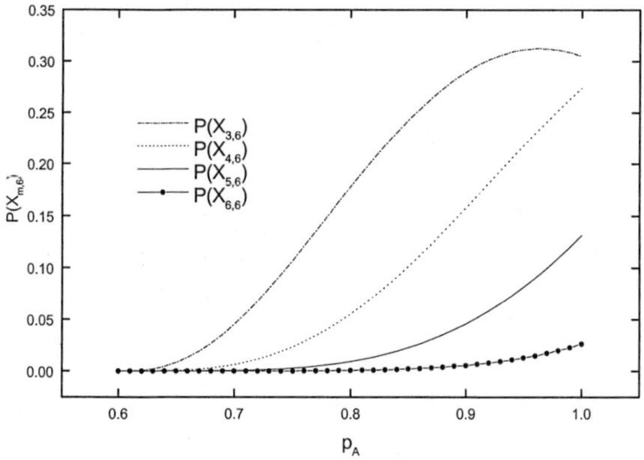

Figure 10. Probability, P($X_{m,6}$), that an amine molecule, belonging to the infinite network, is connected to m segments, as a function of the amine groups conversion, p_A.

Fig.11 shows a plot of μ and ν as a function of p_A, as obtained by eqs. 11 and 12.

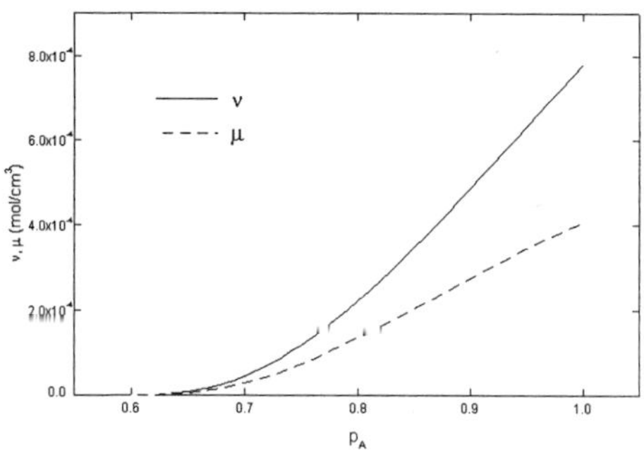

Figure 11. Crosslinking density, μ, and concentration of the active network chains, ν, as a function of the amine groups conversion.

Combining Eqs. 8 through 12 the dependence of L on ν is obtained. As shown in Fig.12 a large part of the curves at every temperature is linear as predicted by the theory of rubber elasticity for the shear modulus. If a linear dependence on ν is assumed for shear, G, and Young, E, moduli then also the bulk modulus K and hence L should be characterized by the same dependence, since:

$$K = \frac{\frac{E}{3}}{3 - \frac{E}{G}} \tag{13}$$

However, as reported by Nielsen[11], the theory of rubber elasticity predicts moduli far too small for extremely high crosslinked materials. Despite of these limitations, in this case a linear correlation between L and the concentration of active network chain is observed during crosslinking as also observed by Nielsen[11].

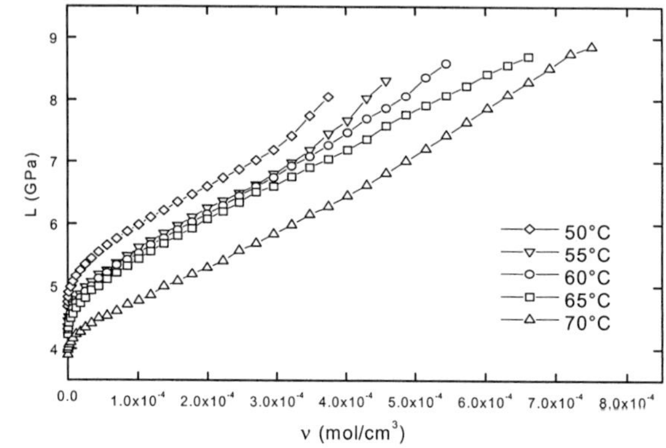

Figure 12. Correlation between L and the concentration of active network chains, ν, after the gel point.

The deviation from linearity close to the gel point in Fig.12 can be the consequence of errors arising both, from ~~of~~ the choice of a theroretical value for α_{gel} and from the assumption of $p_A = \alpha_{DSC}$. A further deviation from linearity is observed at the end of the reaction for isothermal cure at 50, 55 and 60 °C. DSC analysis (Table 1) indicates that at these temperatures α_{max} and T_g are quite lower than 1 and T_{gmax}=91 °C respectively. This indicates that vitrification occurs during reaction leading to a diffusion controlled reaction termination. Therefore the non linearity in the last part of the L vs. ν curves at these three temperatures can be associated to this transition to a glassy state. This transition could be responsible of a faster increase of L with ν.

However, as shown in Fig.12, the linear growth of the modulus with temperature, predicted by the theory of rubber elasticity, is not observed. This may be explained considering that the very small deformation applied by ultrasonic wave propagation involves only energetic intrachain interactions rather than entropic effects associated with macroscopic stretching of the chain segments[12]. In fact the energetic contribution to the elastic response of a rubber, not dependent on the strain, may be dominant at very small strains[12]. This result, in contrast with the observed linear dependence of L on ν, suggests that the entropic contribution to the ultrasonic bulk longitudinal modulus is not relevant. In this case the results of Fig.11 could be explained simply considering that the crosslinking is associated with a significant stiffening of the materials thanks to the introduction of strong constraints to the molecular mobility. α measures each reaction steps either leading to a simple chain extension or to additional crosslinking during the network growth. On the other hand ν, representative only of the development of crosslinking segments, appears to significantly affect the stiffness of the resin. An additional stiffening effects is observed as an increase of L in Fig.11, in correspondence of the vitrification. Therefore the presented results indicates that the ultrasonic longitudinal modulus of the rubbery epoxy shows a linear dependence on this event.

Conclusions

The isothermal crosslinking of a model epoxy resin cured with an exa-functional amine is studied using propagation of longitudinal ultrasonic wave. The ultrasonic velocity and attenuation, measured by a transmission technique, have been applied to the calculation of the bulk longitudinal modulus. The comparison of ultrasonic and DSC data have indicated that the most significant change of the ultrasonic modulus starts close to gelation. Therefore, a correlation between the ultrasonic modulus and the crosslinking density is proposed using the statistical theory of Miller and Macosko. Then, the degree of reaction obtained by DSC and the concentration of active network chains are correlated on the base of the stoichiometry of epoxy and amine monomers. The longitudinal modulus plotted as a function of the concentration of active network chains showed a linear dependence in the rubbery region, while in the last part of the reaction, the transition to the glassy state is reflected in a sudden growth of the storage modulus. However, the longitudinal modulus, for each value of the crosslinking density, decreases with temperature, in contrast with theory of rubber elasticity. This result suggested that the entropic contribution to the ultrasonic bulk longitudinal modulus is not relevant as a consequence of the small deformation involved in the propagation of the ultrasonic waves.

References

[1] D. Lairez, D. Durand, J.R. Emeryl, *Makromol. Chem, Macromol. Symp.* 45 (1991) p. 31

[2] I. Alig, D. Lellinger, K. Nancke, A. Rizos and G. Fytas *J. Appl. Polym. Sci.* 44 (1992) p. 829

[3] M. Matsukawa and I. Nagai, *J. Acoust. Soc. Am.* 99 (1996) p. 2110

[4] A. Maffezzoli, E. Quarta, V.A.M. Luprano, G. Montagna, L. Nicolais *J. Appl. Polym. Sci.*, 73 (1999) p. 1969

[5] C.V. Macosko, D.R. Miller, *Macromolecules*, 9, (1976), p. 199

[6] D.R. Miller, C.W. Macosko, *Macromolecules*, 9, (1976), p. 206

[7] K. Dusek, *Network formation in Curing of Epoxy Resins,* Dusek, 1985

[8] G. Odian, *Principles of polymerization*, McGraw-Hill, New York, 3rd edn, 1991

[9] P.J.Flory *Priciples of polymer chemistry* Cornell University Press, Ithaca N.Y. 1953

[10] J.D. Ferry, *Viscoelastic Properties of Polymers*, John Wiley and Sons, New York, 1980

[11] L.E. Nielsen and R.F. Landel, Mechanical properties of polymers and composites Marcel Dekker N.Y USA (1994)

[12] J.J. Aklonis and W.J. MacKnight *Introduction to polymer viscoelasticity* Wiley Interscience N.Y. (1982)

Numerical Simulation of Interfacial Debonding in Particulate Composites Considering Material Inhomogeneities

Panayiotis Kakavas

Technological Educational Institute of Patras, Greece

*Nick Anifantis**

Mechanical and Aeronautics Engineering Department, University of Patras, Greece

Summary: A comparative study in the framework of large deformations has been conducted to get insight into micro mechanical response of particulate polymeric composites. The propound model enables finite element predictions of displacement response and resulting debonding of a three-phase material subjected to incremental loading conditions. Predictions employ a unit cell of certain shape under specified loading and constraints. The proposed model involves nonlinear material properties and incorporates a strong or weak non-homogeneous interphase region. The interphase toughness is expressed in terms of the adhesion efficiency between the filler and the matrix, though contact conditions are preserved utilizing special contact elements. Interfacial nonlinear spring joint elements regulate debonding initialization, in relation to the imposed failure criterion. Numerical results are presented and discussed for axial tensile and compressive loadings for a variety of values of the imposed parameters.

Introduction

As the demands on strength and durability of today's engineering materials increase, the advantages of combining the properties of two or more materials into a single member have become apparent. However, the combination of material properties can contribute to failure of structures, although it is of particular interest the integrity of junction between the constituents. Considering idealized conditions, these materials often consist of stiff spherical fillers embedded in a

 CCC 1022-1360/00/$ 17.50+.50/0

soft matrix. In order to rationalize the design of reinforced materials it is of primary importance to have a detailed knowledge of stress concentration induced by the applied load. Debonding is the result of overloading in conjunction with nonlinear behavior of the material expressed in terms of damage criteria, like plasticity, fracture, etc. Several experimental, analytical and numerical methods exist in order to investigate the interfacial conditions. In this context, to describe the progressive debonding behavior various researchers utilize finite element method incorporating fracture criteria that are combined with material damage models. Filler bonding results from diffusion or chemical reactions and possibly the filler clamping from residual stresses. In most composites interfaces extend to an interphasial thin layer consisted of one or several materials deposited on the filler like a coating. Experiments and numerical models have demonstrated that a strong interphase is beneficial to the strength, toughness, lifetime and creep resistance.[1] However, a major concern is that the mechanical performance of composites may deteriorate upon being exposed to extreme environmental conditions. The interphase region is susceptible to such attack and thus attention on this must be focused. The concept of strong interphase has been established on polymeric coated filler composites.[2] Filler debonding results in from the deflection of the cracks that initiate in the matrix. The mechanism of debonding in strong interphase composites is different than the one occurring in weak composites. Both of these phenomena are not clear and are unresolved.

The calculation of stress concentration around a particle embedded within a matrix material has been the concern for many years. The mechanisms for craze and shear band formation were previously investigated for small glass beads embedded in polymeric matrices, such as, polystyrene and polycarbonate subjected to uniaxial tension.[3] For perfectly adhering glass bead in polystyrene matrix the crazes are formed near the poles of the reinforcement. At these regions the stress analyses has shown that maximum dilatation and principal stress exists. For perfectly adhering glass bead in a polycarbonate matrix the shear bands form near the surface of the particle at 45^0 from the poles. These are the regions of maximum principal shear stress and maximum distortion strain energy density. In the case, of poor adhesion between the two phases, both craze and shear band formation, along the interface between particle and matrix are preceded by dewetting. At dewetting a curvilinear crack is formed starting at the pole and propagating in the direction of the equator until 60^0 where a craze or shear band originates at the tip of the interfacial crack. A theoretical analysis was given for detachment of an elastomer from a rigid spherical inclusion by a tensile stress applied to the specimen.[4] The catastrophic debonding is predicted to occur at a critical applied stress when the initial debonds is small. Detachment is assumed to start at an already debonded region and take place by grown of this debonded region when the elastic strain energy thereby released in the matrix is greater than the energy required for further debonded. This statement is a direct application of the Griffith's criterion. The amount of strain energy lost from the system as a result of debonding can be evaluated from the difference between the strain energy levels before and after debonding has taken place. The establishment of a unit cell appropriate for debonding simulations in particulates reinforced with spherical fillers is an alternative numerical procedure.[5] The stress-strain state of the cell has been calculated in the framework of the large deformation approach within a wide range of filler volume concentrations. The choice of the form and loading conditions for structural elements is the most important stage in searching for an adequate simulation of particulate composites behavior. The Griffith's approach was used to describe the initiation and propagation of debonds where the postulates of the existence of small precursor debonds and the invariance of debonds energy during crack growth was adopted. It was shown, that the most probable point of debonding is located at the region of the inclusion's pole where maximum strain and hydrostatic extensions are concentrated. The energy required for further matrix detachment has been defined as the rate of change of the energy increment of our system needed to increase the crack area. This energy is a characteristic of the system matrix-inclusion regarded as the cohesive or adhesive tear energy. The only energy source for debonding is the current strain energy accumulated in the matrix. The tear strength of the matrix may be considered as a magnitude that provides the ultimate bond strength. The threshold tear strength of the matrix correlates with the modulus of elasticity of the material.[4] The elastic properties of a three-phase composite made up of spherical particles coated with a mesophase and dispersed in a continuum matrix were computed through a periodic microfield approach.[6] The interfacial debonding processes in Al/Sic metal matrix composites using three dimensional finite element computations on representative cells has been reported elsewhere.[7] The reinforcements are considered as elastic, while an elastoplastic law describes the behavior of the matrix. The concept of interphase was used for modeling the problem of fiber and/or particulate composites.[8-10]

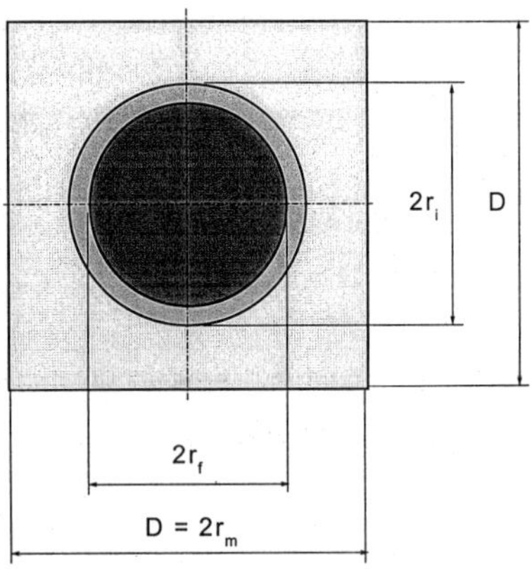

Figure 1. Geometry of the unit cell model.

In an attempt to better understanding the influence of the interphase modulus on the stress transfer characteristics and the debonding process of an axially loaded single spherical particle embedded in a matrix, a computational procedure was developed in the present work. This improved micromechanics analysis investigates the effect of imperfect adhesion and interphase weakness on fracture failure in particulate composite systems. A generalized representative volume element with axially symmetric geometry was chosen, consisting of a spherical filler surrounded by other phases. The matrix and the filler have uniform properties while the interphase have the Young's modulus varying in the radial direction. In order to investigate the influence of the involved parameters to the failure mechanism, a computational analysis has been performed, implementing a regenerative finite element approach, which permits presence of material and strength imperfections. Numerical results illustrate and discuss the evolution of debonding process with load distribution and direction, strength and interfacial adhesion, material inhomogeneities and material damage model.

Non-Homogeneous Interphase

The establishment of a unit cell as unit material configuration appropriate for numerical simulations, based on physical concepts and without enlisting macroscopic notions, may be regarded as a vital problem in today's materials science. In this context, the unit cell considered here is an extension of that proposed in literature.[11] In this cylindrical cell, the spherical filler is embedded on the center of the epoxy matrix. The height of the cylinder is equal to its diameter, i.e., $D = 2r_m$. The radii of the spherical filler and the cylindrical matrix are r_f, r_m, respectively. Around the filler extends an interphasial layer having width $\Delta r = r_i - r_f$ (Figure1). Material properties of the interphase region differ from respective ones of filler and matrix. Physical and chemical reasons require for the interphase properties to comply with those of adjacent phases, enforcing thus strong gradients within this region. Utilizing bulk material properties of filler and matrix in modeling issues of interphase properties[10], the exponential variation in properties of interphase was chosen in the present study for its simplicity and completeness. In order to formulate a generalized model, the filler is assumed to be linearly elastic and the matrix elasto-plastic. The indices f, i and m define material phases of the unit cell.

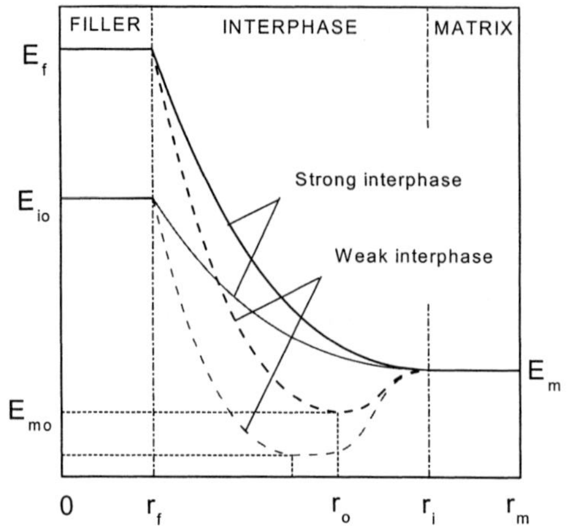

Figure 2. Variation of the modulus of elasticity within the unit cell.

Considering the modulus of elasticity as the dominant material property in stress analysis problems and ignoring Poisson's ratio effects, the variation of the modulus of elasticity within the interphase is shown in Figure 2. An equivalent way to model the degree of debonding between the fiber and matrix lies on the formulation of a simultaneously compliant and imperfect interphase that has lower stiffness than that of the filler and preserves continuity of tractions and displacement at the interphase-matrix interface. This feature of imperfect adhesion is regulated through the ratio $\alpha = E_{io}/E_f$, where E_{io} is the reduced elastic modulus of interphase on $r = r_f$. The upper limit value $\alpha = 1$ describes a compliant three-phase model without jumps on elastic modulus at $r = r_f$, although, the lower limit $\alpha = E_m/E_f$, involves a two-phase model (Figure 2). In this case, the adhesion and subsequently the interphase stiffness decrease to minimum levels, but no material separation or similar type failures occur. When $E_m/E_f \leq \alpha \leq 1$, intermediate three-phase models are established. This general formulation describes imperfect adhesion as an extension of the perfect one, by assigning proper values to the adhesion factor α.

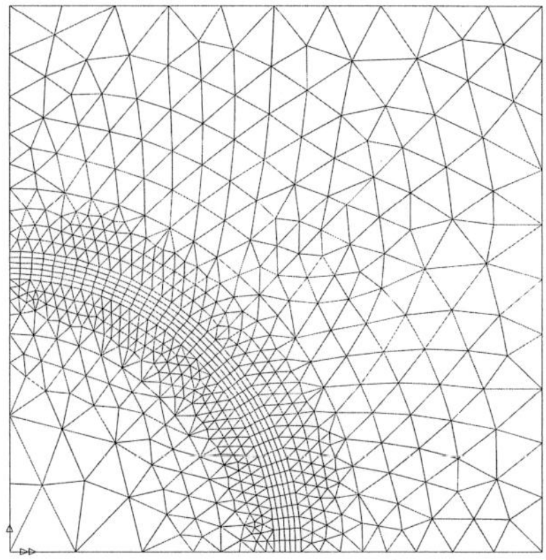

Figure 3. Finite element grid.

The present model involves also a weak interphase by including the presence of impurities, damages or yielding. To a certain extend, this type of weakness is suggested by the cure chemistry of epoxy resin. This type of failure is equivalently simulated by a loss of the interphase stiffness, which may be lower than that of the matrix. Factor $\beta = E_{mo}/E_m$, which varies in the range $0 \leq \beta \leq 1$, express the interphase weakness feature of the model, where E_{mo} is the minimum value of elastic modulus within interphase, occuring on $r = r_o$. Interphase weakness may be affected on third factors like Poisson's ratio variations. Since both the inhomogeneity and stress concentration due to the Poisson ratio are much smaller that those due to the elastic modulus, Poisson effect is neglected here without loss of generality. This general model involves a compliant interphase on the side of the matrix, imperfect on the side of the fiber, and softer than the fiber and matrix in the between. While the first of these features lies on manufacturing reasons, the other two simulate imperfect adhesion and damaged interphase situations. They are independent of each other, may exist or not, and their magnitude is controlled through adhesion and loss stiffness factors, respectively.

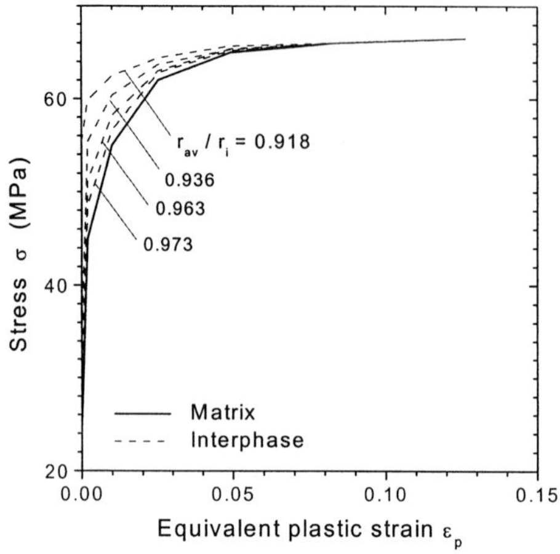

Figure 4. Nonlinear stress-stain behavior of unit cell materials.

Depending on the strength of the intrphase, the modulus of elasticity within the interphase is then assumed to vary, as

$$E_i(r) = E_m\,[1 + (\alpha q - 1)\,R_s(r)] \tag{1}$$

$$E_i(r) = E_m \begin{cases} \beta[1 + (\alpha q - 1)\,R_w(r)], \text{when } r_f \le r \le r_o \\ 1 + (\beta - 1)\,Q(r), \quad \text{when } r_o \le r \le r_i \end{cases} \tag{2}$$

where Equation (1) defines a strong and Equation (2) a weak interphase, respectively. In above equations, the ratio $q = E_f/E_m$, express the overall composite inhomogenuity, and the functions,

$$R_s(r) = \frac{1 - (r/r_i)\exp(1 - r/r_i)}{1 - (r_f/r_i)\exp(1 - r_f/r_i)} \tag{3}$$

$$R_w(r) = \frac{1-(r/r_o)\exp(1-r/r_o)}{1-(r_f/r_o)\exp(1-r_f/r_o)} \quad (4)$$

$$Q(r) = \frac{1}{1+\exp((r-r_c)/c)}, \quad (5)$$

the exponential variation in the transverse direction. The parameter $r_c = (r_o + r_i)/2$, determines the slope of stiffness, c being a constant. Although this model can easily expanded to involve other types of imperfections, in the present paper general spatial inhomogeneities within the interphase, different than those reported above are ignored. Many different interphases can be modeled by varying the parameters α, β, and the type of weakness.

Table 1. Material properties.

Material	E (GPa)	ν	Yield stress (MPa)
Glass filler	70.00	0.220	-
Epoxy matrix	3.50	0.350	25

Table 2. Experimental stress-strain response of epoxy matrix.

Engineering strain (%)	Stress (MPa)
1	25
2	45
3	55
4	62
5	65
6	66
7	66.5

Computational Procedures

While analytical models usually are of simple geometry and boundary conditions, the finite element method employs the same exact mathematical procedure in problem solving and yet possesses the flexibility for complex geometry, inhomogeneous material and boundary conditions modeling. In this study, in order to elucidate the role of unit cell morphology, interphase weakness and imperfect adhesion mechanisms, the deformation, stress response and failure of filler-reinforced composites was numerically investigated. The inherent nonlinearities render the problem unresolved by analytical procedures, leading to numerical approximations. Among these, the finite element method is adopted for its simplicity, generality and ability to treat inhomogeneous nonlinear failure problems effectively. A computational model has been developed that enables stress and failure analysis of any material inhomogeneity by defining the model features by means of appropriate global variables. To reduce the size and the complexity of the procedure, the analysis is performed on a representative unit cell consisted from a spherical filler embedded in a polymeric matrix. Between these phases, extends an inhomogeneous interphase region. A regenerative finite element procedure was developed, which allows solution of any particular composite construction, accounting variations in domain, materials, toughness, weak interphase and imperfect adhesion. Then global design variables include geometric parameters, constituent material properties, and interphase weakness coefficients. Accounting the nodal coordinates for each element lying within the interphase, the elastic moduli are then evaluated and assigned by means of Equations (1) and (2). A computer algorithm was developed in order to evaluate the elastic moduli within the interphase, for the given particular values of the global variables. Within the interphase, material properties then vary in a piecewise manner approaching the exponential variations.

The specificity of the cylindrical unit cell provides a better fit between the modeled and realistic behavior of particulate composites at high filler volume fraction. The radius of the inclusion is a measure of the filler volume fraction. When this radius increases until it touches the lateral surface of the cell, the maximum filler volume fraction is reached. This value is close to that characterizing the ultimate packing of random structures composed of the uniformly sized spheres.[11] The solution of the nonlinear debonding problem has been solved utilizing finite element procedures. Although numerical results are depicted for a particular case, the modeling is quite general and may produce results for any value for filler volume fraction.

(a) (b)

Figure 5. Contours of axial displacement component of the deformed unit cell on, (a) debonding initiation, and (b) fracture propagation.

The finite element mesh adopted for numerical predictions is shown in Figure 3. This model consists of 960 triangular, 200 rectangular and 40 interface, axially symmetric elements, and totally of 15596 nodes. The interphase subdomain has been modeled adopting five layers of axially symmetric rectangular elements. The interface elements are used at filler-interphase interface potential debonding surface to model interphasial failure, and crack initiation and propagation, in an incremental nonlinear analysis. These four-node axially symmetric elements have no geometric properties and connect conjugate nodes of filler surface to corresponding ones of interphase internal surface with nonlinear springs. They have initially zero thickness, but when some relative displacement between the conjugate node pairs and some initiation stress level are simultaneously enabled under loading increase conditions, then failure starts permitting softening of springs followed by large opening displacements. The material toughness is a measure of fracture energy required to debonding initiation process. Depending on the failure process orientation, they describe opening, shearing or complex mode fractures.

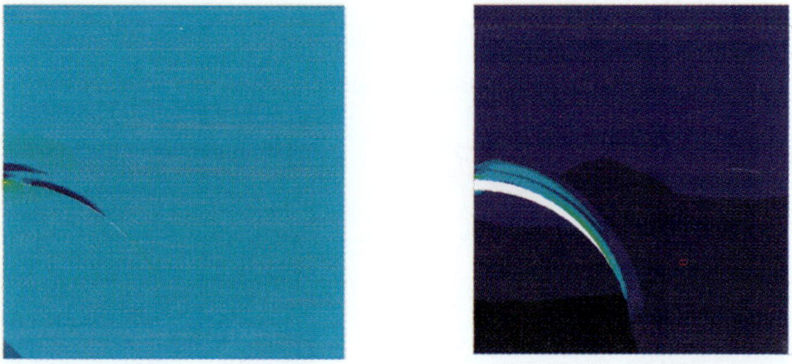

(a) (b)

Figure 6. Contours of radial stress over the deformed unit cell configuration, (a) on the debonding initiation, and (b) on fracture propagation.

Table 1 describes the material properties of the constituent materials. The glass filler is considered as perfectly elastic and the epoxy matrix as nonlinear elastic with experimentally measured stress-strain behavior as that shown in Table 2. To be the material properties of the interphase compliant to those of filler and matrix, it has been assumed that the interphase presents nonlinear elastic behavior with progressive stiffening as the filler is approached. Thus, the stress-strain response of the interphase is a function of radial abscissas. In the case of a strong interphase with $\alpha = 0.85$, Figure 4 illustrates the average values of stress-strain characteristics of interphasial layers. Assuming nonlinear material hardening convention, the equivalent plastic strain ε_{pi} was evaluated with the help of Equations (1-2) and Table 1, by the relation

$$\varepsilon_{pi}=\varepsilon_{pi-1}+\varepsilon_i-\sigma_i/E \qquad (6)$$

where ε_i, σ_i are the strain and stress pairs shown in Table 2, respectively, i being the sequence order. The slopes s_i of the curves in Figure 4, have been accordingly evaluated as follows

$$s_i = E_i/(1 - E_i/E) \qquad (7)$$

In above Equations, $E_i = (\sigma_i - \sigma_{i-1})/(\varepsilon_i - \varepsilon_{i-1})$ is the linearized modulus of elasticity, $E = \sigma_y/\varepsilon_y$ is the conventional modulus of elasticity, $\varepsilon_y = \varepsilon_1$ and $\sigma_y = \sigma_1$ being the yield strain and stress, respectively. Thus, particular nonlinear material models of the interphase may be constructed in terms of the global variables of the problem.

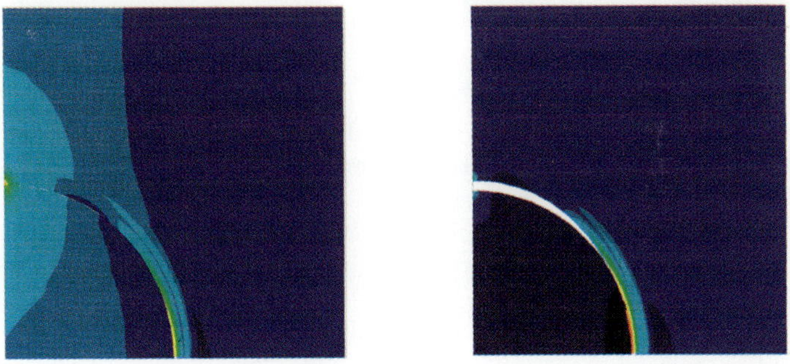

(a) (b)

Figure 7. Contours of axial stress over the deformed unit cell configuration, (a) on the debonding initiation, and (b) on fracture propagation.

The consistensy of the unit cell with the macroscopic behavior of the composite, requires application of a constraint on the lateral surface to keep its shape under deformation. On the top surface of the model, extentional or compressive loading in the form of prescribed vertical displacement it is applied, and on the other sides, usual symmetry conditions are valid. The nonlinear behavior is governed by an elastoplastic constitutive law, which is formulated with a Mohr-Coulomb criterion. Since it is no longer possible to directly obtain a stress distribution, which equilibrates a given set of external loading, an incremental solution procedure has been adopted. At each increment, a linear prediction of the nonlinear response is made, and subsequent iterative corrections are performed in order to restore equilibrium by the elimination of the residual or out of balance forces.

(a) (b)

Figure 8. Contours of shear stress over the deformed unit cell configuration, (a) on the debonding initiation, and (b) on fracture propagation.

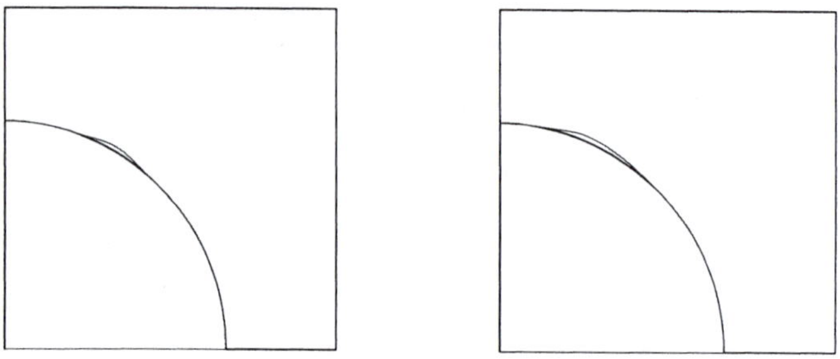

(a) (b)

Figure 9. Sketches showing debonding process evolution under extensional loading on (a) debonding initiation, and (b) fracture propagation stages.

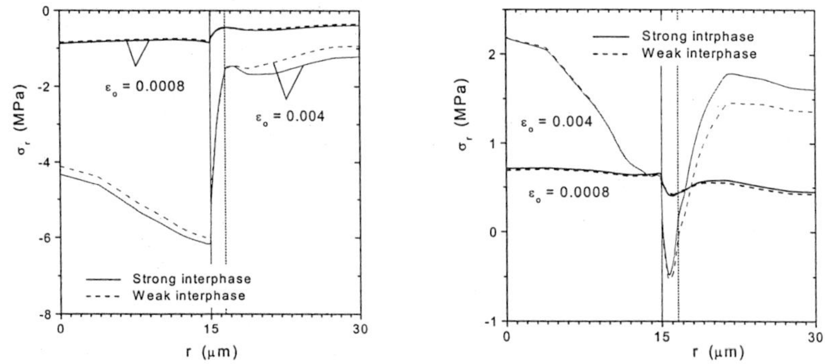

(a) (b)

Figure 10. Variation of radial stress in the direction of the equinoctial plane of the unit cell when is subjected, (a) to tension, and (b) to compression.

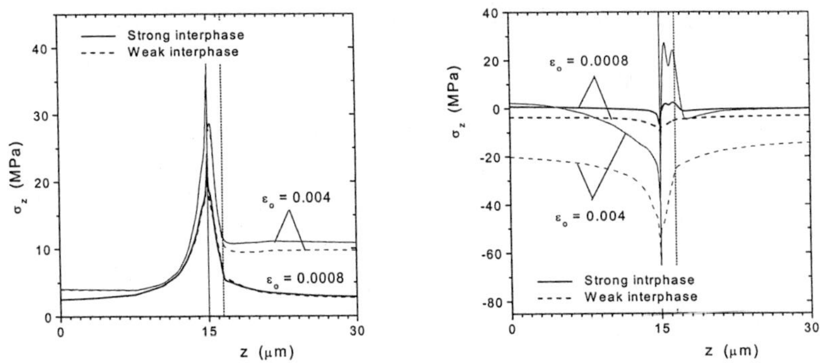

(a) (b)

Figure 11. Variation of axial stress in the direction of axis of symmetry of the unit cell when is subjected, (a) to tension, and (b) to compression.

Numerical Results

In order to investigate the effect of a non-homogeneous interphase on debonding failure process developed within particulates, the computational simulation outlined above has been applied. The particulates used in the present investigation consist of an epoxy matrix reinforced with spherical glass particles. Omitting the long and tedious details, the problem is reduced to several finite element solutions, which yield the desired failure prediction. Varying the global variables of the problem, numerical results were obtained concerning the stress and deformation states, and failure process of particulate composites. To reduce the computational effort, it was assumed that the radius of the filler is $r_f = 15\,\mu m$, the radius of the matrix $r_m = 30\,\mu m$, and the normalized thickness of the interphase $\Delta r_i / r_f = 0.1$. Without loss of generality, numerical results illustrated in the following regard two particular cases. Both incorporate imperfect adhesion conditions between filler and matrix, and presence of strong or weak interphase. Thus, the mechanical properties of the unit cell are such that the composite characterizing parameters have the specific values $q = 28$, $\alpha = 0.85$, and $\beta = 0.5$, for the strong and weak interphase cases, respectively. For the properties of the interface elements, the fracture energy was taken 150 Joule, the initiation stress 65 MPa and the maximum relative displacement $0.001\,\mu m$.[5]

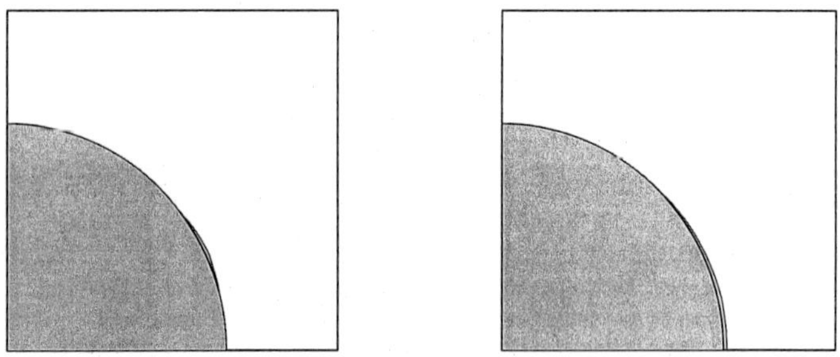

(a) (b)

Figure 12. Sketches showing debonding process evolution under compressional loading on (a) debonding initiation, and (b) fracture propagation stages.

Initially the convergence of the procedure was investigated, considering various mesh refinements, increment magnitudes and relative interfacial displacements. Results shown in the following refer to optimized parameters of the procedure, for which convergence is attained. Based on very dense mesh refinement scheme, propound model does not invlove treatment of field singularities with crack elements, which complicate significantly the numerical procedures. Thus, stress singularities can be only seen as "moving" stress concentrations ahead debonding tips. Considering extensional loading conditions, the two cases are illustrated in the following, regard the applied strains on the top surface of the unit cell $\varepsilon_0 = 0.0005$ and $\varepsilon_0 = 0.004$, which correspond to debonding initiation and some fracture extension, respectively. To be the results more clear, an exaggeration factor of 20 was applied to the deformed shapes. For comparison reasons, the outline of undeformed original shape of the unit cell is drawn also by solid lines. A color key scale of 20 colors was adopted, red and blue being the maximum and minimum values, respectively, while other colors indicate intermediate values. Figure 5 depicts the contours of the axial displacement component. The location of the debonding initiation lies about 30° lower than the pole of the particle and is identical to that experienced experimentally. From this point a crack initiates and propagates simultaneously to the pole and equator positions. When the region of the pole debonds, a depth open crack is formed. The deformations of the particle are very small while within the interphase step deformation gradients develop. Figures 6.a and 6.b illustrate the contours of radial stress at the same loading increments, which correspond to debond initiation and at some crack depth, respectively. It is obvious the stress concentration in the region of interphase and especially in the locus of debond initiation. The fracture propagation is shown to release stresses and fracture surfaces become traction free. Figures 7 and 8 illustrate the contours of axial and shear stresses, respectively, on the same loading increments. A detailed examination of these Figures explains the locus position of initial debonding. Where some relative displacement occurs and simultaneously some high stress is developed, there debonding starts. The rate of crack propagation in both tips depends on both the stress concentration and the relative deformations produced on increase of loading (see Figure 9).

When the unit cell is subjected to extensional or compression loading, Figures 9 and 10 depict a comparison of between radial and axial stress distributions, respectively. Radial stress distribution is close to the equator ($z \approx 0$), although axial stress distribution lies on the axis of symmetry ($r \approx 0$). Numerical results are plotted for two absolute values of the applied strain on the top surface of the cell, i.e., $\varepsilon_o = 0.0008$ and $\varepsilon_o = 0.004$, which correspond to debonding initiation and to the development of some fracture, respectively. In all cases shown, imperfect adhesion is valid with $\alpha = 0.85$. This value has been chosen close to perfect case ($\alpha = 1.0$), in order to study the sensitivity of the debonding process to small changes in the involved parameters. The weak interphase regards a situation where $\beta = 0.5$. Interphase weakness a little affects the stress distribution. In the axial direction, at some time instants high stress concentration occurs that does not yield debonding because the relative displacement is small enough. Generally speaking, the extensional loading yields more intense stress fields than the corresponding compressive. Under compressive loading, the debonding process is different than that observed under tensional loading (see Figure 12). In the former case the filler behaves like a solid body, the debonding starts on the side of the equator and very small crack opening deformations occur. When tensile or compressive loading is applied, Sketches 9, 12 and Figures 10, 11 explain somewhat the difference in the debonding mechanism as affected by the particular shape of unit cell and imposed lateral constrains. At the tips of the debonding surface, which is axially an interfacial crack, high stress intensification occurs featuring the principles of fracture mechanics.

Conclusion

A generalized computational model has been developed for the prediction of the debonding characteristics of particulate composites incorporating a heterogeneous interphase region. This model enables simulations of both imperfect adhesion and interphasial weakness by proper formulation of the interphase properties. Then, numerical studies were conducted to demonstrate how the interphase strength and the adhesion efficiency affect the local deformation and stress states and so forth the fracture mechanisms. The loci of high stress concentration and step stress gradients are investigated and inter-related to failure initiation and propagation process. Although the interphase plays a crucial role in the development of load transfer mechanisms, the presence of material imperfections within this region intensifies stress gradients and causes stress relaxation of the material structure. Present model is general and may be used for further analysis and hence provide a basis for further understanding of the toughening mechanism for fiber bridging.

[1] B. T. Lauke, T. Schuller, W. Beckert, *Computational Materials Science* **2000**, 18, 362.
[2] F. J. Rebillat, J. Lamon, A. Guette, *Acta Materiallia* **2000**, 48, 4609.
[3] M. E. J. Dekkers, D. Heikens, *Journal of Materials Science* **1985**, 20, 3865.
[4] A. Gent A, *Journal of Material Sciences* **1980**, 15, 2884.
[5] V. V. Moshev, L. L. Kozhevnikova, *Inter. Journal of Solids and Structures* **2000**, 37, 1079.
[6] J. M. Llorca, M. Elices, Y. Termonia, *Acta Materiallia* **2000**, 48, 4589.
[7] H. Haddadi, C. Teodosiu, *Computational Materials Science* **1999**, 16, 315.
[8] J.L. Chaboche, R. Girard, A. Schaff, *Computational Mechanics* **1997**, 20, 3.
[9] J.Y Li, *International Journal of Solis and Structures* **2000**, 37, 5579.
[10] C. C. Kiritsi, N. K. Anifantis, *Computational Materials Science* **2001**, 20, 86.
[11] J.D. Bernal, G. Mason, *Nature* **1960**, 4754.

Macromol. Symp. **180,** *109–123 (2002)*

Physical & Thermal Properties of Model Polysiloxane Rubbers: Impact of Crosslink Density and Tin Concentration

M. Patel, P.R. Morrell and A.R. Skinner*

MLP group, AWE plc, Aldermaston, RG7 4PR, U.K.

Summary: The influence of the tin octanoate catalyst on the physical and thermal properties of RTV 5370 polysiloxane rubbers has been studied. To assess the likely influence of crosslinking on a number of physical and thermal properties in polysiloxanes, "model siloxane networks" (representing networks of well defined composition/structure) have been formulated by the hydrosilylation of polysiloxane diols of known average molecular weight with tetraethoxysilane curing agent. It was found that linear swell and the crystallisation melting transitions of these systems were both significantly affected by changes in crosslink density. A selection of RTV5370 foamed rubbers with different tin concentrations were prepared in a similar manner to assess the influence of the tin catalyst. For these materials it was found that the area of the crystalline melting transition decreased with increasing tin concentration, an effect indicative of increased crosslinking. Samples with double the standard amount of tin (10% wt catalyst) show a shift in the crystallisation transition to higher temperature with no further effect beyond this concentration. Furthermore, the storage modulus (E') of the rubbers at room temperature appears to be independent of the tin catalyst concentration. Through correlation with the data from the model siloxanes, these results suggests that increasing the catalyst concentration appears to induce additional crosslinking interactions that are stable at low temperatures but not stable at room temperature.

Introduction

In general, there are two main technologies used to produce cellular polysiloxane rubbers, each having certain advantages and disadvantages over the other. The first method is based on the use of low molecular weight silicone gum in which the cellular structure is produced by a simultaneous tin catalyst accelerated condensation polymerisation reaction and gas formation within the mould. Normally a dihydroxy terminated poly dimethylsiloxane containing a small amount of silica filler, together with a polyfunctional (f>2) crosslinking agent (e.g. tetrapropylsilicate) is used. A suitable quantity of organotin catalyst is added to initiate the reaction. Tin (II) and Tin (IV) catalysts (in particular dialkyltindicarboxylates or tindicarboxylates) are most frequently used to catalyse the condensation reaction of the dihydroxy terminated

 CCC 1022-1360/00/$ 17.50+.50/0

polydimethylsiloxane with the alkoxysilanes. For a polyfunctional crosslinking agent this will result in gelation[1]. Initiated by the tin catalyst, the reaction of terminal or side-chain silane hydrogens with the hydroxy groups on the polysiloxane chains will generate hydrogen gas as the blowing agent in these systems. The second method of producing polysiloxane foamed rubbers requires the use of relatively high molecular weight gum in which water soluble spherical particles are added (as pore formers) and then subsequently removed (by washing with water) after the rubber has been shaped and set in a mould.

In an assessment of the effects of stannous 2-ethylhexanoate catalyst on the polysiloxane diol in the absence of crosslinking agent, a possible chain extension reaction arising from the catalyst induced condensation of the silanols has been reported by X.W.He[2]. Based on these findings, it was concluded that excess residual catalyst will result in effectively longer polysiloxane diol chains and as a result, the molecular weight between crosslinks (M_c) formed from end-linking in the presence of crosslinking agent (e.g. tetraethoxysilane) will be significantly higher than expected. Similar studies by S.J. Clarson[3] showed that, at concentrations of 1 and 1.5 wt %, the stannous 2-ethylhexanoate promotes significant chain extension of the polysiloxane diol chains in the absence of the crosslinking agent at 20°C. For catalyst concentrations of <0.1% weight catalyst, however, no significant chain extension effects were observed in the polysiloxane gum. Chain extension effects were reported to be insignificant in the presence of the curing agent.

Research into the mechanism by which tin compounds catalyse the condensation reaction has shown the system to be complex with the actual catalyst being an organotin hydroxide which is a hydrolysis product of the tin octanoate[4]. The formation of the reactive tin hydroxide is dependent on finite levels of water within the gum. Carboxylic acid is also produced during the hydrolysis reaction which is consistent with the observation that the cure is retarded by the addition of carboxylic acid to this equilibrium reaction[5]. Subsequent reaction of the organotin hydroxide with the alkoxy silane is reported to generate silanolates (compounds containing Sn-O-Si linkages that are sensitive to silanols). The organotin hydroxide, being the true catalyst, is regenerated as the polymerisation proceeds.

The aim of the work reported here is to develop an improved understanding of the influence of the tin catalyst on the physical and thermal properties of the material. Of particular interest is whether there is a link between rubber crosslink density and the catalyst concentration. Due to the relatively complex nature of commercially available polysiloxane rubbers and the inherent disadvantages of the condensation polymerisation technique (e.g. operator dependence, batch to batch variation, difficulty controlling density, cell size, product modulus, etc.), "model siloxane networks" (representing networks of well defined composition/structure) have been formulated and characterised. These model compounds, of differing crosslink density, were prepared so that the influence of crosslinking on a number of physical and thermal properties including linear toluene swell, crystallisation and mechanical performance could be assessed. A method for the synthesis of model compounds utilising the end-linking of functionally terminated linear polydimethylsiloxane chains has been reported by R.C. Hedden[6] and was used to generate the systems studied in this work.

Experimental Procedures

Materials & synthesis of model polymers

The RTV5370 polysiloxane rubber samples of differing tin content were prepared by adding a known amount of XY-70 catalyst (stannous 2-ethylhexanoate) to 30g of RTV5370 gum (supplied by Rhone Poulenc and containing the curing agent tetrapropyl orthosilicate) and the mixture was vigorously stirred for approx. 15 seconds. 20g of this mixture was added to a 15cm diameter (2 mm deep) mould. The polymer was cured under ambient conditions for 20 minutes and then postcured at 70°C for 16 hours. A selection of RTV5370 polysiloxane rubbers with catalyst concentrations of 4.2, 5,5.8,6.7,10 and 15 wt % were prepared using this method.

The dihydroxy terminated polydimethylsiloxanes used in these investigations were purchased from Fluorochem Ltd. A total of six polymers were obtained having quoted M_n (polystyrene equivalent number average molecular weights) of 400-700, 1500-2000, 4200, 18000, 26000 and 58000 g/mol. Networks were prepared by a hydrosilylation end-linking reaction at room temperature. This involved stirring the dihydroxy terminated polydimethylsiloxanes with the stannous 2-ethylhexanoate catalyst (1 wt%) and the tetraethoxysilane (5 wt %) curing agent. The samples were left to cure overnight under ambient conditions. The soluble fractions of the networks were extracted in toluene to remove residual impurities and to extract out unattached chains.

Characterisation of the model siloxanes

Chemical Analysis

The dihydroxy terminated polydimethylsiloxane gums were analysed using ^{1}H, ^{13}C and ^{29}Si Nuclear Magnetic Resonance (solution NMR). For each polymer, the results confirmed the presence of two different proton environments, namely CH_3 and OH, and the ^{29}Si NMR showed the presence of two silicon environments, namely Si-OH (-11ppm) and O-Si-O (-22ppm). The ^{13}C NMR studies also showed the presence of the methyl groups.

Thermal Analysis

A TA Instruments 2910 Differential Scanning Calorimeter (DSC) was used to assess the thermal characteristics, including phase transitions, within the model polymers over the temperature range -140°C to 100°C.

Thermomechanical Analysis

A TA Instruments 2940 Thermomechanical Analyser (TMA) was used to monitor the linear swell response (with respect to thickness) of the model samples. Liquid toluene was injected into the chamber containing a sample of known thickness. The dimensional changes were monitored over time until equilibrium swell conditions were achieved.

Solvent Uptake Measurements

A known amount of rubber was placed in toluene for 48 hours and then re-weighed to determine the solvent uptake. The crosslink density (expressed as the G-factor) was calculated from the solvent uptake[7].

Characterisation of the RTV5370 polysiloxane rubber

Tin Content

The RTV5370 polysiloxane rubber samples were analysed for tin content by Neutron Activation Analysis (NAA). These measurements were conducted at the NAA facility of Imperial College at Silwood Park. On analysis, the RTV5370 polysiloxane rubbers prepared using 4.2, 5,5.8,6.7,10 and 15 wt % of tin octanoate catalyst, were found to have a tin concentration of 1.1, 1.4, 1.5, 1.9, 2.9 and 3.9 wt % respectively. The tin concentration was also re-determined after extraction in toluene to assess the amount of tin tightly bound to the network.

Soluble Fraction Measurements

Extractable matter measurements were made on the RTV5370 polysiloxane rubber samples. A known amount of rubber was placed in toluene and extracted for 48 hours (ambient conditions). After extraction, the samples were allowed to dry overnight before being re-weighed. The percentage solvent extractable matter was calculated from the change in sample weight.

Differential Scanning Calorimeter

A TA Instruments 2910 Differential Scanning Calorimeter (DSC) was used to assess the phase transitions within the RTV5370 polymers. Sample were heated at 10°C/min.

Thermomechanical Analysis

A TA Instruments 2940 Thermomechanical Analyser (TMA) was used to assess the load-bearing properties of the samples. Experiments were conducted to monitor the stress relaxation characteristics of the RTV5370 polysiloxane rubber samples at 25°C using a constant compressive strain of 10%.

Dynamic Mechanical Thermal Analysis (DMTA)

A polymer Labs DMTA was used to assess the dynamic mechanical properties of the RTV5370 polymers. These measurements were performed by the Department of Chemistry, Physics and Environmental Sciences at the University of Sussex.

Results and Discussion

Model polymers of defined structure

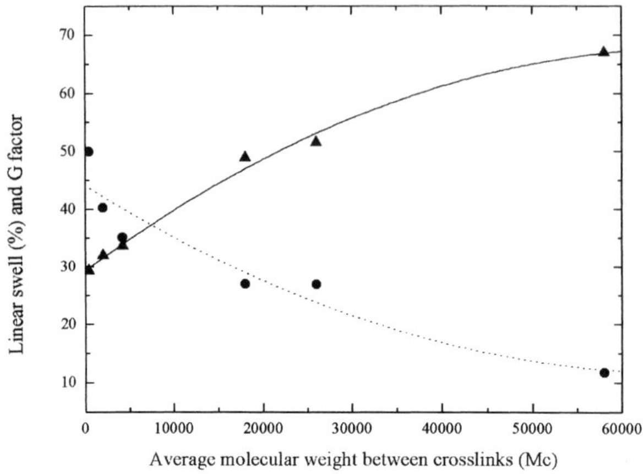

Figure 1: Linear swell (▲) and solvent uptake (G-factor, •)

Figure 1 shows the variation in linear swell characteristics for toluene uptake with crosslink density. It is evident that the model polymers clearly show differences in network crosslink density with the most highly crosslinked material (Mc, 400 g/mol) showing the lowest linear swell (30% increase in sample thickness) and the highest G-factor value (50%). G-factors are determined from the solvent uptake and are representative of the crosslink density of the rubber as described by Flory[8].

These results demonstrate that the hydrosilylation end-linking method of capping precursor chains of different molecular weights was successful in preparing structures that vary in network crosslink density and which do not suffer significantly from network imperfections (such as unattached chains).

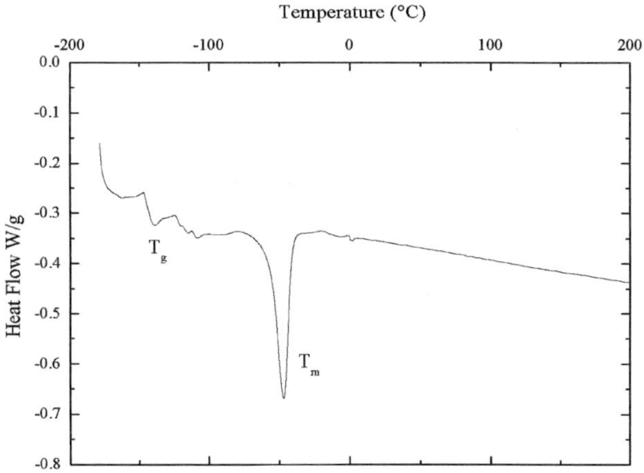

Figure 2: Typical DSC trace of model siloxane (Mc: 26,000 g/mol)

In order to assess the impact of crosslink density on thermal characteristics (in particular thermal transitions) the model polymers were tested using Differential Scanning Calorimeter (DSC). Figure 2 shows a typical DSC trace of a model polymer. The main thermal transitions are a glass transition temperature (T_g) at –120°C and a melting transition (T_m) centred at –45°C. The DSC trace does not give a clear indication of a heat of crystallisation (T_c) presumably because the rate of crystallisation and the cooling rate is such that the crystallites form during the cooling setup of the cycle (and are therefore not recorded by the DSC) with little crystallisation on heating.

Our studies indicate that the network crosslink density can significantly influence the crystallisation characteristics of the polymer, see Figure 3. Interestingly, within experimental error, no significant changes in Tg were observed as the crosslink density was varied.

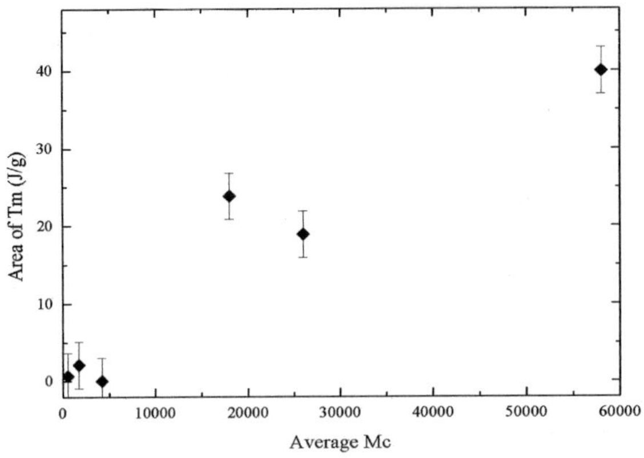

Figure 3: Impact of crosslink density on crystallinity

Crystallisation is a phenomenon strongly dependent on the ability of polymer chains to align/move into required positions and it is presumably more difficult for a polymer with a high crosslink to carryout this process. This effect could be considered similar to the impact of bulky side groups (e.g. phenyl rings) which will also reduce the ability of a polymer to crystallise.

RTV5370 polysiloxane rubber samples of different tin content

To assess the extent of incorporation of the catalyst into the polymer network, toluene extractable matter measurements were made on the RTV5370 polysiloxane rubber samples and the extracted samples re-analysed for tin using NAA. Samples with catalyst concentrations of 5.8 wt% (representing a Tin concentration of 1.5%) and above showed larger toluene soluble fractions (see Figure 4) and a reduction in tin content (see Table 1), suggesting that some unreacted catalyst is being leeched out by the extraction process.

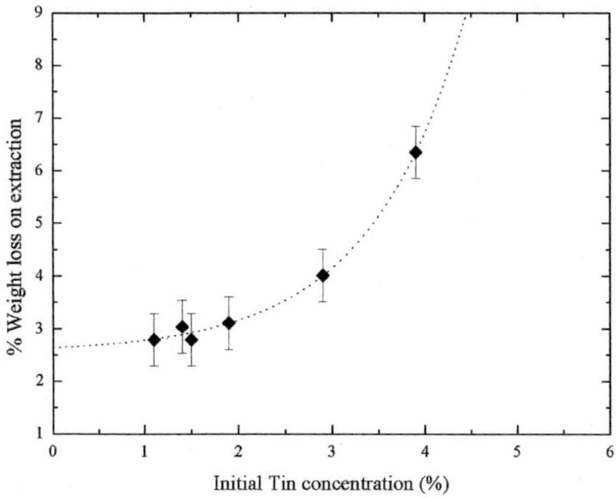

Figure 4: Solvent extractable matter

The % weight loss on extraction correlates with the weight loss expected from the decease in % tin (arising from loss of residual unreacted catalyst) if organic species associated with the catalyst are also taken into account. The presence of additional catalyst does not appear to be causing significant changes to the amount of non-network siloxane species present within the rubber. For those samples containing 10% and 15% XY70 catalyst, a significant amount of tin still remains after solvent extraction suggesting it is tightly bound to the network.

Catalyst Concentration (%)	4.2	5.0	5.8	6.7	10.0	15.0
% Tin*	1.1	1.4	1.5	1.9	2.9	3.9
% Tin after toluene extraction**	1.2	1.4	1.5	1.8	2.4	3.0
change	-0.1	0.0	0.0	0.1	0.5	0.9

Table 1: Tin content (from NAA) of toluene extracted samples (* ± 0.05%, ** ± 0.07%)

Figure 5 shows a typical DSC trace of a RTV5370 polymer. Whilst the overall features are similar to those found for the model polymer (see Figure 2) in this case, the crystallisation melting transition is significantly lower in intensity and there is a large heat of crystallisation (centred at –85°C). The reduced size of the melting peak is presumably because RTV5370 polysiloxane is formulated by the manufacturers to contain phenyl groups to help enhance high temperature stability and reduce crystallinity at low temperatures. The large heat of crystallisation (T_c) may be explained by the rate of crystallisation and the cooling rate being such that the crystallites predominately develop during the heating stage of the cycle with little crystallisation on cooling i.e. rapid cooling 'quenches' the material which maintains its amorphous character even at low temperatures.

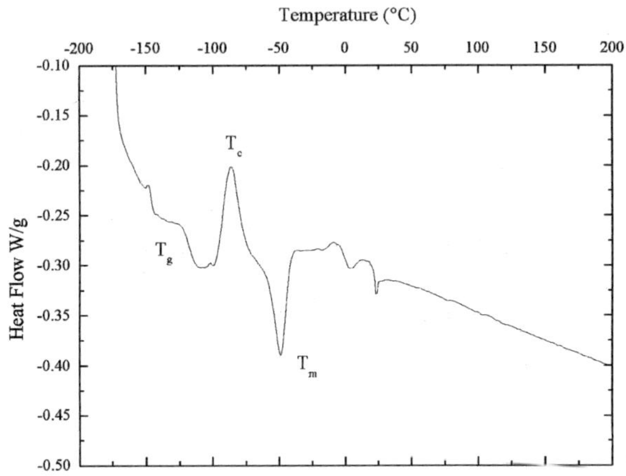

Figure 5: Typical DSC Trace of virgin RTV5370 rubber

Figure 6 shows that increasing the catalyst concentration is clearly detrimental to the crystallisation properties with samples having higher catalyst levels showing a very small or no melting (T_m) transition.

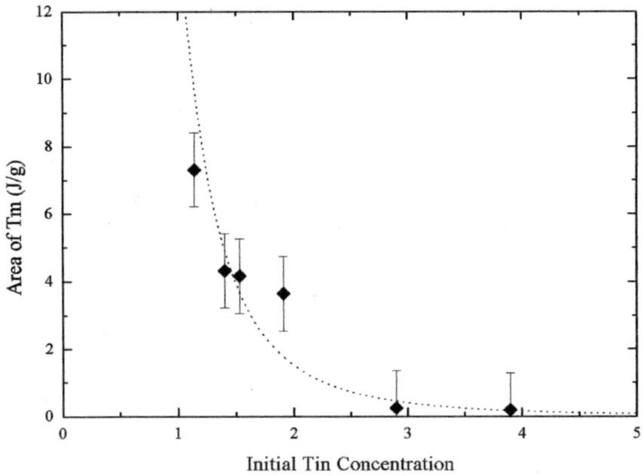

Figure 6: Impact of tin concentration on crystallinity

Evidently, the additional catalyst is reducing the ability of the polymer chains to align sufficiently for crystallisation and comparison with the results for the model siloxanes, suggests this is due to increased crosslinking of the polymer.

To examine whether changes in crosslink density are taking place with increasing catalyst content, the mechanical behaviour of the RTV5370 rubbers was assessed by DMTA. The tan δ plots (see Figure 7), show that increasing the catalyst concentration leads to a material that requires more energy to align the chains for crystallisation (indicative of increased crosslinking) as reflected in a small but measurable shift of the main tan δ peak from –95 to –80°C as the catalyst concentration is increased to 10 wt %. Above this the effect is negligible. The area under the main tan δ peaks also differ indicating differences in the degree of crystallinity. In the corresponding modulus plots (see Figure 8) the RTV5370 rubbers show an increase in E' above Tg followed by a falling modulus. The increase in E' is representative of crystallisation of the rubber, followed by melting. Samples with high tin concentration, in particular the material containing 10 wt % catalyst, show a shift to higher temperature with no further effect beyond this concentration and this observation is consistent with the tan δ data.

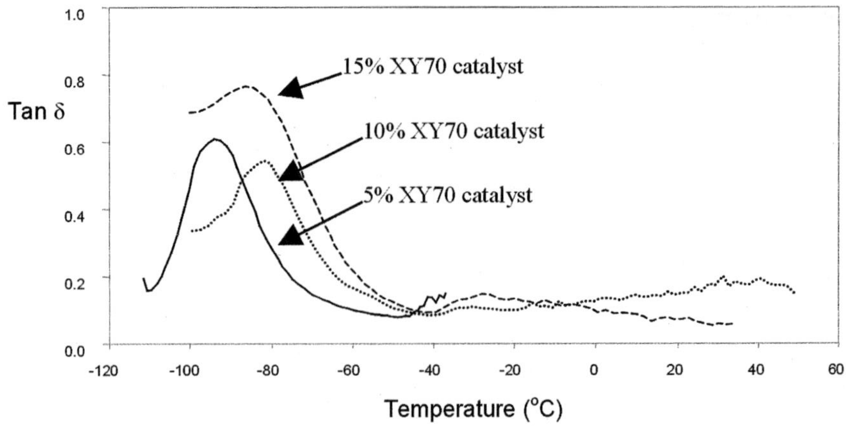

Figure 7: DMTA plot of tan delta (damping)

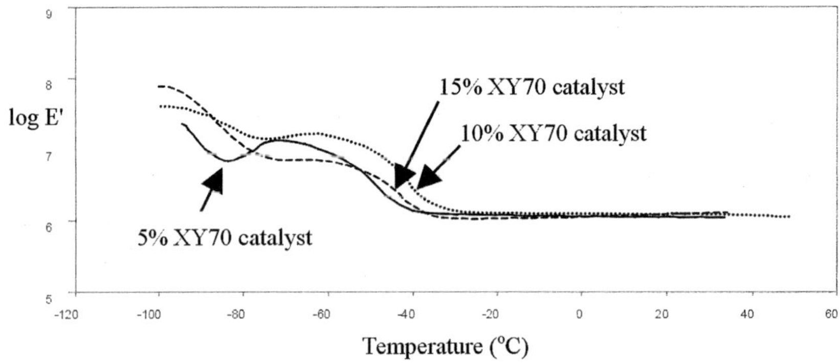

Figure 8: DMTA plot of storage modulus

Comparison of DMTA and DSC results

The DMTA and the DSC results agree in that the main effect of the tin catalyst appears to be to influence the low temperature crystallisation properties of the rubber, presumably by increasing crosslink density. The DMTA data suggests that for samples having double the standard amount of tin, the crystallisation behaviour shifts to higher temperatures (indicating greater energy is required to induce crystallinity as a result of increased crosslinking) and for tin concentrations above this no further change is indicated. The DSC results show that higher catalyst concentrations result in very small or no melting (T_m) transitions. This effect possibly arises from increased crosslinking in the polymer and the reduction in the ability of the polymer chains to align sufficiently for crystallisation.

A possible source of increased crosslinking resulting from increasing the tin concentration is the formation of tin(IV) oxide (SnO_2) from hydrolysis of the catalyst. Such species, having active surface groups (e.g. OH), might increase the effective crosslink density of the polymer[9] by forming hydrogen bonding like interactions. Figure 9 shows an overall increase in the rate of stress relaxation (suggesting a more mobile structure) for those samples having a higher tin concentration. This suggests that any possible increased crosslinking from the tin is weak/unstable in terms of load bearing/retention characteristics at room temperature. Catalyst hydrolysis effects will also generate increased quantities of carboxylic acid within the polymer. Such acidic species will accelerate siloxane exchange reactions and also lead to an increased rate of stress relaxation[10] as the tin concentration is increased.

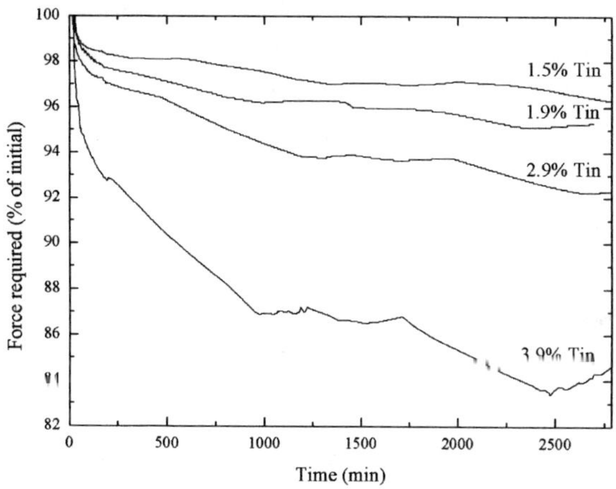

Figure 9: Stress relaxation properties (10% strain and 25°C)

Conclusions

The model siloxanes of different crosslink density have proved a useful tool in developing our understanding of the link between structure and physical/thermal properties. Linear swell, as well as the crystallisation melting transition, was both found to be influenced by changes in crosslink density.

Studies on the effect of tin catalyst concentration in RTV5370 polysiloxane rubbers has shown that the area of the crystalline melting transition, as measured by DSC, decreased with increasing tin concentration. DMTA studies have shown that increasing the catalyst concentration leads to a shift of the tan δ and storage modulus (E') peaks to higher temperatures. It is also found that the storage modulus of the rubbers at room temperature appears to be independent of the tin catalyst concentration. Through correlation with model siloxanes, these results suggests that increasing catalyst concentrations appears to induce additional crosslinking interactions that are stable at low temperatures but not stable at room temperature. The stability of these bonding interactions also appears to be low when placed under load for long periods of time, as evidenced by poorer stress relaxation properties as the tin concentration is increased. A possible source of increased crosslinking, through hydrogen bonding, is the formation of stannic oxide particles from hydrolysis of the catalyst.

Acknowledgements

The Authors are grateful to Professor Norman Billingham & Miss Amiya Chaudhry, from the Department of Chemistry, Physics and Environmental Sciences at the University of Sussex for the DMTA results. We would also like to thank Dr. Julian Murphy for helpful discussions and Dr. Mark Read for helping with the figures.

References

[1] J.WEN, J.E. MARK, J.J. FITZGERALD, Macromoleculer reports, A31, 1994, p429.
[2] X.W.HE, J.M.WIDMAIER, J.E. HERZ AND G.C. MEYER, Eur. Polym. J., 24, p1145, 1988.
[3] S.J. CLARSON, ZAIMING WANG AND J.E. MARK, Eur. Polym.J. vol. 26, No. 6, p621, 1990.
[4] F.W. VAN DER WEIJ, Makromol. Chem., 181 (12), 1980, p2541
[5] J.TOYNBEE, Polymer, 35, No. 2, 1994, p438.0
[6] R.C. HEDDEN, H. SAXENA, AND C.COHEN, Macromolecules, 2000, 33, p8676
[7] ZHANJIE TAN, RAIMUND JAEGER, G. JULIUS VANCSO, Polymer, 35, No 15, 1994, p3230
[8] P.J. FLORY, "Principles of Polymer Chemistry", Cornell University Press,
[9] P.G. HARRISON, "The Chemistry of Tin", Blackie, 1989.
[10] M.ZELDIN, J. Polym. Sci. 1983, 21, p1361.

Macromol. Symp. **180,** *125–131 (2002)*

Preparation and Characterization of Iodinated Poly(vinyl alcohol) Microfibril

Won Seok Lyoo[1], Sam Soo Kim[1], Han Do Ghim[2], Jae Pil Kim[2], Ick Chan Kwon[3], Chul Joo Lee[4]*

[1]School of Textiles, Yeungnam University, Kyongsan 712-749, Korea
[2]School of Materials Science and Engineering, Seoul National University, Seoul 151-742, Korea
[3]Biomedical Research Center, Korea Institute of Science and Technology, PO Box 131 Cheongryang, Seoul, Korea
[4]Electronic Materials and Devices Research Center, Korea Institute of Science and Technology, PO Box 131 Cheongryang, Seoul, Korea

Summary: Iodination of poly(vinyl alcohol) (PVA) microfibril, which was obtained from saponification of poly(vinyl pivalate), was conducted before and after zone drawing at various conditions. The resulting PVA microfibrils were characterized by differential scanning calorimeter and scanning electron microscopy. Surface morphologies of these PVA microfibrils showed some differences between PVA microfibril iodinated after and before drawing. Crude shapes of PVA microfibrils iodinated after drawing indicated that iodine decreased the structural regularity severely. On the other hand, PVA microfibrils iodinated before drawing showed relatively ordered surfaces. This was ascribed to the enhanced molecular ordering of PVA microfibrils due on zone drawing. Iodinated PVA microfibrils showed a decrease in crystal melting temperature of about 100 ^{o}C compared to the untreated sample. PVA microfibrils drawn after iodination showed relatively higher crystal melting temperature than those of microfibrils iodinated after drawing. These results were considered as the proofs of the changes in crystalline lattice of the PVA microfibrils. Effects of drawing temperature on sublimation of iodine were also evaluated.

Introduction

Poly(vinyl alcohol) (PVA) obtained by the saponification of poly(vinyl ester) is a linear semicrystalline polymer, which has been widely used as fiber. PVA fibers have high tensile and compressive strengths, tensile modulus, abrasion resistance due to its highest crystalline lattice modulus, and superior biocompatibility.[1-3] Recently, Lyoo *et al.* found that a PVA fiber of well-oriented microfibrillar structure was formed during saponifying poly(vinyl pivalate) (PVPi) to PVA.[4-10] These PVA microfibrils show the appearance and characteristics of natural fibers, such as cotton and jute. PVA microfibrils show more enhanced mechanical properties than spun fibers. Furthermore, PVA microfibrils have abundant microvoids, which are suspected to improve the biocompatibility.

Embolization is the operational therapy of controlling the flow of blood to the specific vein or portion of it to prohibit the supply of nutrition to some kind of cell, such as cancer cell. Another kind of embolization is physical treatment for blood circulation, such as the aneurysm surgery.[11-13] Cancer cell-killing embolic application has been paid attention to for its possibility and effectiveness for replacement of commercialized metallic coil shaped embolic materials; metallic materials have some problems during operation for its sharpness of cut face. PVA microfibrils, in contrast, show superior properties, including enhanced biocompatibility and less risk of scratching, but hard to be detected by X-ray during the operation because of its lack of radio-opacity. Radio-opacity is very important characteristics for embolic materials for the minute positioning control of materials in human organ.

Iodine is reported to reform the crystalline lattice of PVA. Changes in the crystalline lattice of PVA microfibrils by iodine sorption indicated that iodine molecules intruded into crystalline, as well as amorphous regions, to form new substitution types of crystalline lattice.[14,15] It is valuable to relate the radio-opacity introduced and property changes in operational point of view. In this study, iodination of PVA microfibrils was conducted with varying iodination conditions. For the consideration of the effects of crystalline structure on the formation of iodine-complex, PVA microfibrils were drawn before and after iodination at various conditions. The resulting PVA microfibrils were characterized by differential scanning calorimeter and scanning electron micrograph. Iodinated PVA microfibrils showed a decrease in melting temperature about 100 ^{o}C. Crystalline lattice change and physical and chemical characteristic of iodinated PVA microfibrils were also evaluated according to the amount of iodine sorption.

Experimental

Materials

PVA microfibrils were prepared by saponification of PVPi having number averaged degree of polymerization (P_n) of 25,000. Resulting PVA microfibril had P_n and syndiotactic diad content of 10,000 and 61.0%, respectively. Iodine and potassium iodide (KI) were purchased from Aldrich Co. Extra-pure grade reagents were used without further purification. Water used for all the procedures was deionized.

Iodination of PVA microfibril

Aqueous iodine solution of $4x10^{-1}$ mol/l was prepared by mixing of I_2 and KI at mole ratio of 1 to 2. Drawn and untreated PVA microfibrils were iodinated in this aqueous iodine solution for 48 hours at room temperature. Iodinated PVA microfibrils were washed in water for 1 hour at room temperature to wash off the adhesive iodine. Iodinated PVA microfibrils, then, were dried at room temperature without vacuum to prevent the sublimation of iodine.

Zone drawing of PVA microfibril

Iodinated and untreated PVA microfibrils were drawn by zone drawing method. Drawing temperatures were changed from 100 to 200 ^{o}C at fixed drawing rate and weight. Draw ratio was measured as the length gain and ranged from 2 to 4 times.

Desorption of iodine

Iodinated PVA microfibrils were treated with water for 2 hours at 40 ^{o}C. Amount of water was controlled to 1,000ml/1g of PVA microfibrils. Finally, PVA microfibrils were dried at room temperature in atmosphere.

Characterization

Appearances of PVA microfibrils were observed by scanning electron microscopy (SEM). Thermal characteristics of PVA microfibrils were evaluated by using differential scanning calorimeter (DSC).

Results and Discussion

Zone drawing, due to short residence time within a heated drawing zone, has many advantages such as: fewer possibilities of back folding of molecular chain, of microcrystallite formation, of thermal degradation, and of energy loss.[16-23] Zone drawing condition, therefore, can affect amount of iodine sorption of PVA microfibrils. Fig. 1 shows the iodine adsorption of zone drawn PVA microfibril according to drawing temperature and resulting draw ratio. Iodine adsorptions were decreased with increasing zone drawing temperature. Decreases in iodine adsorption can be explained as the increases of regularities of PVA microfibrils by drawing. In addition, effect of draw ratio on iodine adsorption showed the expected tendency; PVA microfibril of higher draw ratio, and resulting higher degree of molecular regularity, showed lower iodine adsorption. Especially, PVA microfibril drawn 4 times at 200 °C showed steeply decreased amount of adsorption. It is well known that crystalline structure of PVA can be penetrated by iodine. In this study, it is assumed that drawing can effectively prevent the penetration of PVA crystalline structure by iodine.

Iodine has high degree of sublimation. The heat on drawing, therefore, evaporated the iodine resulting in weight loss of iodinated PVA microfibrils. Weight losses of iodinated PVA microfibril (weight gain for iodination 39.29%) versus drawing temperature and resulting draw ratio were shown in Fig. 2. Weight losses were calculated with respect to absorbed iodine. Increase in drawing temperature results in the increase of the weight loss of iodinated PVA microfibril. PVA microfibrils drawn 4 times show much higher weight loss than those of drawn twice at same temperature. It is ascribed that hot zone duration time of PVA microfibril drawn 4 times is much more than that of drawn twice.

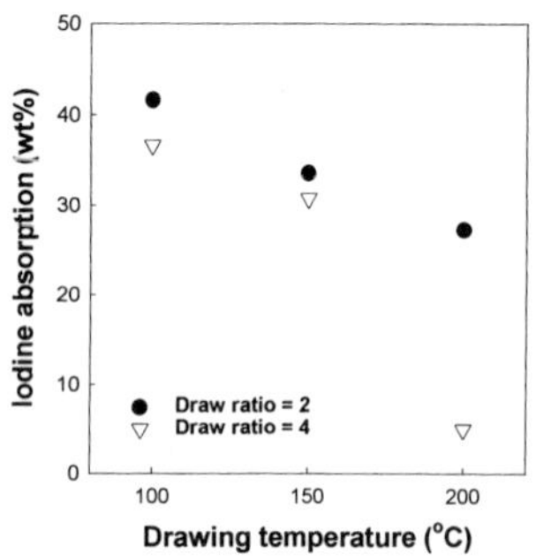

Fig. 1. Iodine adsorption of PVA micro-fibril according to drawing temperature. (Draw ratio were controlled to 2-4)

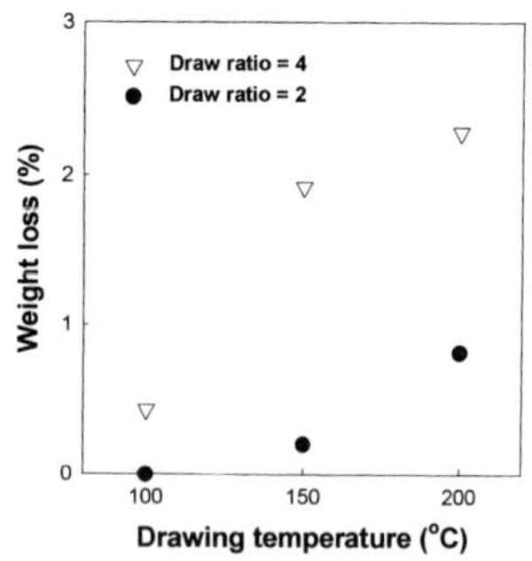

Fig. 2. Weight loss of iodinated PVA microfibril vs. drawing temperature. (Draw ratio were controlled to 2-4)

Iodine desorption of PVA was an important subject for its usage as a polarized film.[24,25] To use iodinated PVA microfibrils as embolic materials, water resistance of iodine desorption should be improved as well. (a) and (b) of Fig. 3 show the amounts of iodine desorption of PVA microfibrils drawn before and after iodination, respectively. Iodine desorption can be depressed by increasing drawing temperature at both cases. Iodine desorption of pre-drawn PVA microfibrils, however, decreased much faster. But this cannot be a clue of fastness of iodine because PVA microfibril drawn 4 times at 200 °C absorbed so less iodine. PVA microfibrils iodinated after drawing show relatively stable iodine desorption values of *ca.* 20%. This result can be related to the penetration depth of iodine. PVA microfibril drawn before iodination should have tighter internal structure than microfibrils drawn after iodination at iodination stage. Therefore, iodine cannot penetrate to the depth enough for drawn PVA microfibrils. This loose-penetrated iodine can be evaporated much more easily.

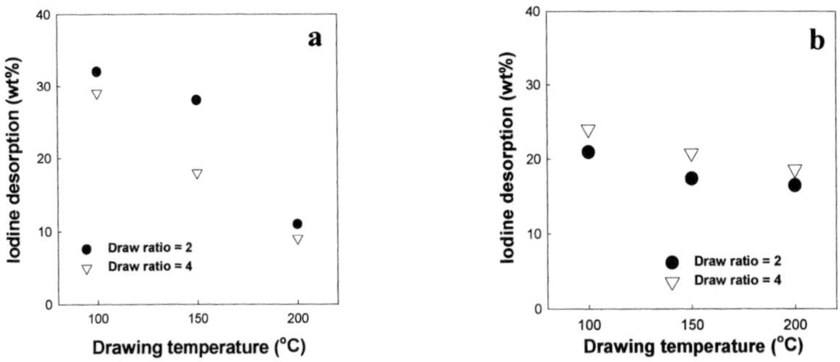

Fig. 3 Iodine desorption of PVA microfibrils versus drawing temperature; (a) drawn before iodination, (b) drawn after iodination. (2 hours in water at 40 °C)

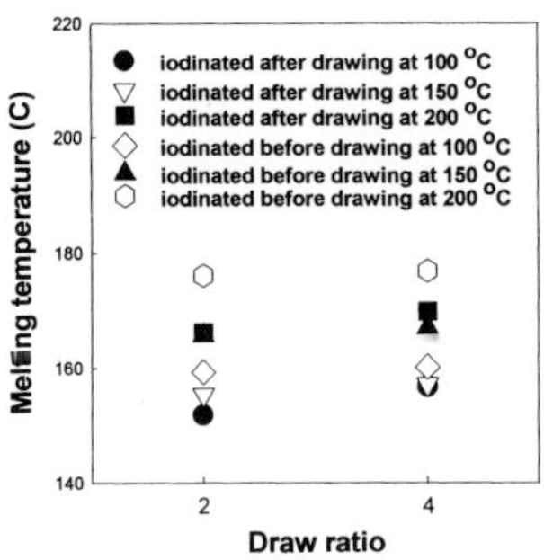

Fig. 4. Crystalline melting temperatures of PVA microfibrils versus draw ratio.

Changes of crystalline structure can be indicated by crystalline melting temperature (T_m). Fig. 4 shows T_m of PVA microfibrils drawn before and after iodination. Temperature was scanned from room temperature to 300 °C at 10 °C/min of heating rate. T_m of untreated PVA microfibril is about 260 °C.[4] On the other hand, PVA microfibrils of this study show (T_m)s of ranging from 150 °C to 180 °C. These changes can be attributed to the reduction in structural regularity by the penetration of iodine.

Surface morphologies of iodinated PVA microfibril were shown in Fig 5. Iodinated PVA fibrils after zone drawing showed much crude surface than fibrils iodinated before drawingn. From this result, it can be confirmed that iodine can also penetrate the PVA crystalline oriented by drawing effectively.

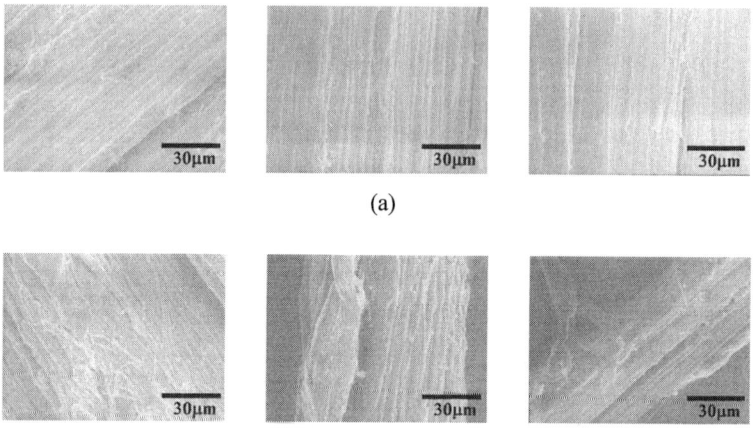

Fig. 5. SEM photographs of PVA fibrils; (a) drawn after iodination, (b) drawn before iodination (x1,000).

Conclusion

In this study, iodination and zone drawing of PVA microfibrils were conducted at various conditions. Iodine adsorption of undrawn PVA microfibril reaches to 39.29% and iodine evaporations in following zone drawing were ranging from 0% to 2.3 wt% of iodinated PVA microfibrils. Weight loss by iodine evaporation increased with the increase in drawing temperature and this tendency was enlarged for PVA microfibrils of higher draw ratio. Iodination of drawn PVA microfibril was also performed and the amount of iodine adsorption reduced from 42% to 5% with increasing drawing temperature. Iodinated PVA microfibrils were steeped in water at 40 °C for 2 hours. PVA microfibrils iodinated after drawing showed drastic decrease in iodine desorption ranging from 33% to 9% with increasing in drawing temperature. However, PVA microfibrils iodinated before drawing showed the value of about 20% for all drawing temperature. (T_m)s of PVA microfibrils increased with increasing drawing temperature and draw ratio, ranging from 150 °C to 180 °C. But these values were about 100 °C lower than those of untreated PVA microfibrils, *ca.* 260 °C.

Financial support provided by Ministry of Health & Welfare (HMP-99-Gr-2-049) in Korea is gratefully acknowledged.

[1] K. Toyoshima, in "*Polyvinyl Alcohols*", C. A. Finch, Ed., John Wiley and Sons, New York 1973, pp. 339-388.
[2] I. Sakurada, in "Polyvinyl Alcohol Fibers", M. Lewin, Ed., Marcel Dekker, New York 1985, pp. 3-9 and pp. 361-386.
[3] M. Masuda, in "Polyvinyl Alcohol-Development", C. A. Finch, Ed., John Wiley and Sons, New York 1991 pp. 403-422 and p. 711.
[4] W. S. Lyoo, W. S. Ha, *Polymer* **1996**, *37*, 3121.
[5] W. S. Lyoo, W. S. Ha, *J. Polym. Sci., Polym. Chem. Ed.* **1997**, *35*, 55.
[6] W. S. Lyoo, J. Blackwell, H. D. Ghim, *Macromolecules* **1998**, *31*, 4253.
[7] W. S. Lyoo, S. N. Chvalun, H. D. Ghim, J. P. Kim, J. Blackwell, *Macromolecules* **2001**, *34*, 2615.
[8] W. S. Lyoo, J. H. Kim, J. H. Choi, B. C. Kim, J. Blackwell, *Macromolecules* **2001**, *34*, 3982.
[9] W. S. Lyoo, B. J. Kim, W. S. Ha, *J. Korean Fiber Soc.* **1996**, *33*, 231.
[10] W. S. Lyoo, J. H. Yeum, H. D. Ghim, B. C. Ji, W. S. Yoon, J. P. Kim, *J. Korean Fiber Soc.* **2000**, *37*, 487.
[11] V. P. Chuang, S. Wallace, *Radiology* **1980**, *135*, 2959.
[12] V. P. Chuang, C. S. Soo, S. Wallace, *AJR* **1981**, *136*, 729.
[13] H. Demachi, O. Matsui, T. Takashima, *Cardiovasc. Intervent. Radiol.* **1991**, *14*, 158.
[14] W. O. Hermann, W. Haehnel, *Ber. Dtsch. Chem. Ges.* **1927**, *60*, 1658.
[15] H. Staudinger, K. Frey, W. Starck, *Ber. Dtsch. Chem. Ges.* **1927**, *60*, 1782.
[16] S. S. Han, W. S. Yoon, W. S. Lyoo, C. J. Lee, B. C. Ji, *J. Macromol. Sci.-Phys.* **1997**, *B36*, 1.
[17] S. S. Han, W. S. Yoon, J. H. Choi, S. Y. Kim, B. C. Ji, W. S. Lyoo, *J. Appl. Polym. Sci.* **1997**, *66*, 1583.
[18] B. C. Ji, G. S. Jeong, W. S. Yoon, S. S. Han, W. S. Lyoo, *Polymer Testing* **1999**, *18*, 299.
[19] W. S. Lyoo, S. S. Han, W. S. Yoon, B. C. Ji, J. Lee, Y. W. Cho, J. H. Choi, W. S. Ha, *J. Appl. Polym. Sci.* **2000**, *77*, 123.
[20] B. C. Ji, E. K. Kim, C. J. Lee, W. S. Lyoo, W. S. Yoon, S. S. Han, *Polymer(Korea)* **1995**, *19*, 846.
[21] W. S. Lyoo, S. S. Han, J. H. Choi, Y. W. Cho, W. S. Ha, *J. Korean Fiber Soc.* **1995**, *32*, 1023.
[22] B. C. Ji, W. S. Yoon, S. Y. Kim, *J. Korean Fiber Soc.* **1993**, *30*, 328.
[23] S. S. Han, S. Y. Kim, *J. Korean Fiber Soc.* **1994**, *31*, 1012.
[24] W. S. Lyoo, J. H. Yeum, J. H. Choi, B. C. Ji, H. D. Ghim, J. P. Kim, T. H. Noh, W. J. Yoon, *Polymer Testing* **2001**, *20*, 503.
[25] W. S. Lyoo, J. H. Yeum, J. H. Choi, H. C. Song, B. C. Ji, J. P. Kim, T. H. Noh, W. J. Yoon, T. S. Cheong, *J. Appl. Polym. Sci.* **2001**, *82*, 108.

Macromol. Symp. **180,** *133–140 (2002)*

Recycling of a Starch-Based Biodegradable Polymer

F.P. La Mantia, R. Scaffaro, C. Bastioli†

Dipartimento di Ingegneria Chimica dei Processi e dei Materiali, Università di Palermo, Viale delle Scienze, 90128 Palermo Italy
Tel: + 39 0916567203, Fax: + 39 0916567280, email: lamantia@dicpm.unipa.it

†Novamont, Via Fauser 8, 28100 Novara
Tel: + 39 0321699.602, Fax: + 39 09321699602, email: Bastioli@materbi.it

Summary: A new starch-based polymeric system, ZI1OU from Novamont, mainly composed of starch and polycaprolactone, was reprocessed several times in an extruder to investigate the recyclability of this biodegradable polymer. A previous investigation of the thermomechanical degradation in a mixer has been also done. The degradation is mostly due to the thermal stress but the presence of the mechanical stress strongly increases the degradation kinetic. During melt processing two concurrent processes take place: the first is the degradation, i.e. the breaking and shortening of polymeric chains, mostly occurring in the PCL phase; the second is the formation of some crosslinked structure in the starch phase. The rheological and mechanical properties, measured on samples subjected to repetitive extrusion operations carried out in a single screw extruder, clearly indicate that, in the adopted processing conditions, no significant modifications are observed. Only after five extrusions some decrease of these properties was measured.

Introduction

Synthetic polymers are used in many technological applications and their use is continuously growing. Unfortunately, among many great advantages , there are also some properties, like the low density and the high ageing resistance, that can give a serious environmental impact. This is the reason why in the last years a "plastic waste science" has grown up to try to minimise the negative effects of the presence of plastic materials in the environment at the end of their life cycle. Nowadays two main paths, trying to solve this problem, can be individuated. The first is recycling, that is either recovery of the material for a further "re-use" (primary and secondary recycling), monomers or feedstock recovery (tertiary recycling) or energy recovery (quaternary) [1-5]. The second possibility - complementary, not alternative at all - is the use of biodegradable plastics; these are polymer systems that can be by bacteria which are able to oxidize them to transform the materials into carbon dioxide, water and biomass [6-7].

 CCC 1022-1360/00/$ 17.50+.50/0

For these peculiar characteristics, to enhance the economic of these materials and to further reduce the environmental impact, the mechanical recycling of these biodegradable polymers is particularly convenient, but only if the recycled materials show properties that are not poor. A previous work present in literature[8] deals with the recycling of blends of polyethylene and starch i.e. a partial biodegradable material.

The aim of this work is to study the properties of a commercial starch based biodegradable polymer after multiple processing steps in an extruder, simulating recycling and processing operations. In particular, a completely biodegradable material will be used.

Experimental Part

The biodegradable polymer used for this work is an injection moulding grade of MaterBi series Z (ZIO1U) manufactured and kindly supplied by Novamont. It is constituted for about 50% of corn starch and for the remaining of polycaprolactone, a synthetic, biodegradable aliphatic polyester. The polycaprolactone (PCL) is a high molecular weight polymer commercially known as PCL 878 manufactured by Union Carbide and has been characterized for comparison. Like other similar systems [9], this material is characterized by an interpenetrated structure where low molecular weight organic compounds break the H-bonding avoiding the formation of a cross-linked material. This sample is completely biodegraded in 45 days following ASTM 5338.

Before processing, the material has been dried in order to avoid hydrolytic scission. The processing apparatus is a single screw extruder (Brabender) with a thermal profile of 100-130-170-170 °C and a rotational speed of 100 rpm. Under these conditions the estimated residence time is about 90 sec. Degradation tests were done in a batch mixer (Brabender Plasticorder) under different processing conditions. This temperature has been chosen as it is only slightly higher than the typical processing temperatures that are in the range 130 °C – 150 °C.

The rheological measurements were carried out using a CEAST, (Italy), constant rate capillary viscometer, mod. Rheoscope 1000. All of the tests were performed at 170 °C with a capillary having 1 mm diameter and length-to-diameter ratio (L/D) equal to 40. The Bagley's correction was not applied because of the large length-to-diameter ratio, while the Rabinowitsch correction was applied throughout.

The tensile properties were evaluated with an Instron equipment mod. 1122 on specimens (thickness about 0.50 mm, width 10 mm) cut out from compression moulded sheets, prepared by using a Carver laboratory press at 170 °C.

The morphology was studied by scanning electron microscopy (SEM) using a Philips mod. 501 apparatus. The samples were fractured in liquid nitrogen and coated with gold using an SPI sputter coater. Calorimetric tests were performed using DSC7, Perkin Elmer.

Solubility tests and intrinsic viscosity measurements have been carried out by using chloroform (for the solubilization of PCL) and dimethyl solfoxide (for the solubilization of starch) at the temperature of 30 °C. The separation has been carried out in two steps, firstly solubilizing the PCL, and then solubilizing the residual precipitate constituted by starch.

Discussion of the Results

Before analysing the behaviour of the material after recycling operations it is useful to study its thermomechanical degradation as a function of the most important processing parameters. The three parameters that, together with structure and chemical nature of the polymer, influence the degradation of the materials during the processing operations are temperature, mechanical stresses and residence time inside the apparatus.

In Fig. 1 the flow curves of samples processed in the closed mixer under different processing conditions are reported.

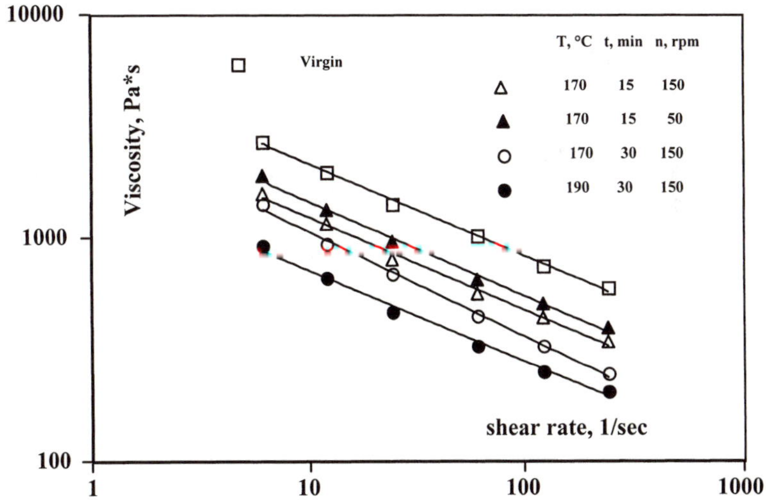

Fig. 1: Flow curves of samples processed in the mixer in different processing conditions.

As expected, the viscosity of the material decreases with increasing time, temperature and also by increasing the rotational speed which increase the mechanical stress applied to the melt.

The above data clearly indicate that some degradation occurs during these severe processing operations. However, it is not possible to discriminate if the degradation occurs because of a decrease in the molecular weight or if the decrease of the viscosity is also due to some change in the morphology, for example due to some reduction in the compatibility, as occurred, because of degradation, in other blends [10]. This last hypothesis, however, is not confirmed for this interpenetrated system, by the SEM observations. Indeed , the morphology does not change as a consequence of the melt processing undergone in these conditions. A reduction in ~~of~~ the molecular weight, as a result of the thermomechanical stress, is then responsible for the decrease of the viscosity. This last observation, together with the fact that most part of the degradation occurs at low processing times, suggests that this is a thermomechanical degradation. The degradation rate is, indeed, decreasing with time and with decreasing the molecular weight. It is worth to remark that the mechanical degradation increases with stress and then with the viscosity of the melt. However, it is not possible to discriminate which of the two components of this multicomponent system is more prone to degrade

The dimensionless viscosity (at 170 °C and at a shear rate of 12 s^{-1}) relative to samples of ZIO1U and of pure polycaprolactone processed in different conditions are reported in Table1. The dimensionless viscosity is obtained by dividing the value of the processed sample by that of the same virgin material.

Table 1. Dimensionless viscosity of PCL and ZIO1U processed in different conditions.

Processing conditions	PCL	ZI1OU
T = 170 °C, n = 50 rpm, t = 15 min	0.80	0.65
T = 170 °C, n = 150 rpm, t = 15 min	0.70	0.51

The degradation level of the multicomponent system seems higher than that of the pure synthetic component. This means that the starch degrades faster or that the degradation of PCL is accelerated by the presence or by the degradation products of the starch, as observed in other multicomponent systems [11]. The decrease of the molecular weight of the PCL is confirmed by the intrinsic viscosity measurements reported in Table 2.

Table 2. Intrinsic viscosity of PCL and starch in virgin and recycled ZIO1U.

Sample	η_{in} PCL, dl/g	η_{in} starch, dl/g	Remarks
ZIO1U virgin	0.761	1.110	
ZIO1U (150 rpm, 5 min)	0.731	1.236	partially insoluble
ZIO1U (150 rpm, 30 min)	0.721	1.165	partially insoluble

On the contrary, the intrinsic viscosity of the amorphous starch phase, reported in the same table, increases and some insoluble material is also observed.

In Fig. 2 the viscosity of recycled samples up to five extrusions has been reported. It is worth noting that for the first three passages in the extruder no significant variation of the viscosity can be detected and only after the fourth passage there is a certain reduction in the viscosity especially at low shear rates, suggesting that for low processing times this materials can be reprocessed several times.

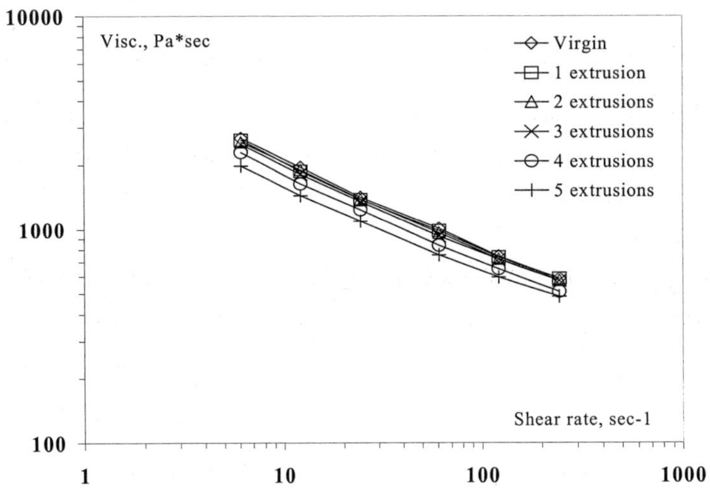

Fig. 2: Viscosity of MaterBi processed several times.

This behaviour is quite different from the one observed during the degradation tests where a remarkable degradation level was reached. In that case the tests were carried out in particularly severe conditions especially for the processing times that were no less than 15 minutes. Actually, in the extruder the average residence time is about one minute and the results indicate that during this time no important degradation effects occur. In the extrusion conditions, even after five passages the viscosity is quite similar to the one of the virgin materials. The rheological results are confirmed by the mechanical tests. In Fig. 3 the tensile stress, the elongation at break and the impact strength are reported for the virgin and for the recycled materials. Both tensile stress and elongation at break are not substantially worsened with respect to the virgin material and only a small reduction of the impact strength can be observed. However, the only property that seems to be slightly affected by the re-processing step is the tensile modulus. The viscosity at 12 s^{-1}, the elastic modulus and the heat of fusion (H_f) of the PCL phase for a specimen of virgin polymer and of the same material after 3 and 5 passages in the extruder are reported in Table 3. The heat of fusion was obtained by the analysis of the calorimetric curves which are not reported here.

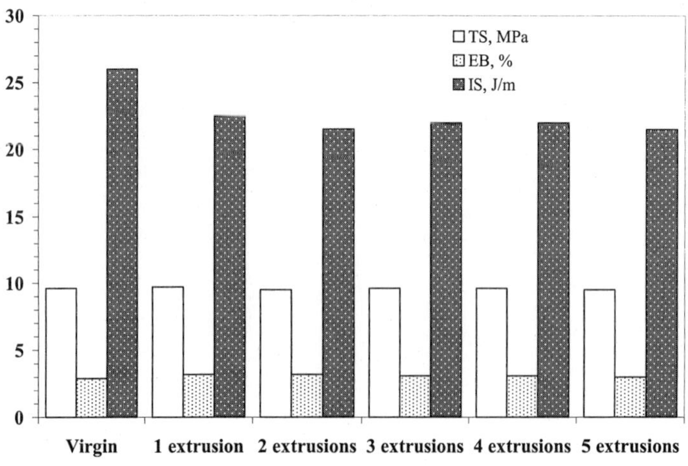

Fig. 3: Tensile stress, elongation at break and impact strength of virgin and recycled material.

As previously said, the viscosity shows relevant changes only after five reprocessing steps but on the contrary, the modulus is increasing. This behaviour was already observed in some semicrystalline polymers [12-14] and not in amorphous polymers[14]. It can be attributed to the increase of crystallinity due to the reduction of molecular weight of PCL, as demonstrated by the decreasing of the viscosity and confirmed by the increasing of the heat of fusion of the PCL semicrystalline phase .

Table 3. Viscosity and mechanical properties of virgin and recycled ZIO1U.

Sample	Viscosity, Pa*s	Modulus, MPa	H_f, J/g
ZIO1U virgin	1950	570	53
ZIO1U (3 passages)	1895	611	54.1
ZIO1U (5 passages)	1415	680	55.9

Considering the previous results, the decrease in the viscosity is mostly due to the reduction of molecular weight of the PCL, while the rise of the modulus to both the presence of small amount of crosslinked molecules in the starch phase and to the rise of crystallinity in the PCL phase.

It can be concluded that during melt processing two concurrent processes take place: the first is the degradation i.e. the breaking and shortening of polymeric chains, mostly occurring in the PCL phase; the second is the formation of some crosslinked structure in the starch phase. The first phenomenon is responsible for the decrease of viscosity and for ~~of~~ the slight increase of crystallinity, while the second is the responsible for ~~of~~ the increase of the modulus, property highly sensitive even to small changes in the polymeric structure.

The above data clearly demonstrate that, for this starch based biodegradable polymer, the conditions of temperature, mechanical stresses and residence time typical of extrusion operations allow to obtain reprocessed materials with properties similar to those of the virgin material. Such behaviour in not typical for all the recycled polymers that, due to the very high degradation rate, cannot often be re-used in the same processing apparatus and show very poor mechanical properties. It must be added that this class of polymer (MaterBi Serie Z) contains some plasticizers but the variation of the properties cannot be reasonably attributed to any change of structure of the molecules either to any interaction with the macromolecules because these compounds are chemically inert, thermally stable and lowly volatile at the processing temperatures used in this work.

Conclusions

The starch based biodegradable polymer ZIO1U is relatively stable to thermal degradation and to the applied mechanical stress that can accelerate the degradation kinetic. The degradation is mostly connected with chain breaking of the synthetic component molecules and with the prevalent crosslinking of the starch phase. Nevertheless, the typical reprocessing conditions does not influence in a relevant way both rheological and mechanical properties. The material can be then reprocessed in the same apparatus used for the virgin polymers and for the same applications.

Acknowledgment

This work has been financially supported by Università di Palermo, RS 60%

[1] Leidner, J., *Plastics Waste*, M. Dekker, New York, 1981

[2] Ehrig, R.J., *Plastics Recycling*, Hanser, Munich, 1992

[3] La Mantia, F.P., ed, *Recycling of Plastic Materials,* Chemtec, Toronto, 1993

[4] La Mantia, F.P., ed, *Recycling of PVC and Mixed Plastics*, Chemtec, Toronto, 1995

[4] Brandrup, J., Bittner, M., Michaeli, W., Menges, G., *Recycling and Recovery of Plastics*, Hanser, Munich, 1996

[5] Sheirs, J., *Polymer Recycling*, Wiley, New York, 1998

[6] Griffin, G.J.L., ed, *Biodegradable Polymers*, Blackie Academic & Professional, London 1994

[7] Scott, G., Gilead, D., eds, *Degradable Polymers*, Chapman & Hall, London 1995

[8] C.R. Armistead in Griffin, G.J.L., ed, *Biodegradable Polymers*, Blackie Academic & Professional, Ch. 5, pages 97-115, London 1994

[9] Bastioli, C. in Scott, G., Gilead, D., eds, *Degradable Polymers*, Chapman & Hall, London 1995

[9] La Mantia, F.P., Valenza, A., Bernardo, C.A., Covas J.A., Kozlowski, M., Steller, R., Bryjak, M., "*The Polymer Processing Society Europe/Africa Regional meeting*", Gothenburgh, August 19-21, 1997 Sweden, pages 7p3-7p4

[11] La Mantia, F.P., in *Handbook of Polymer Degradation* Hamid, S.H, Amin, M.B., Maadhah, A.G., eds, M. Dekker, New York 1992

[12] La Mantia, F.P., Vinci, M., Polym. Deg. Stab., **45**, 121 (1994)

[13] Marrone, M., La Mantia, F.P., Polymer Recycling, **2**, 17 (1996)

[14] La Mantia, F.P. *Frontiers in the Science and Technology of Polymer Recycling* (ed. G. Akovali, C. A. Bernardo, J. Leidner, L. A. Utracki, M. Xanthos), p. 249-269, Kluwer Academic Publishers, Dordrecht (1998)

Macromol. Symp. ***180,*** *141–151 (2002)*

Scavenging of Halogen in Recycling of Halogen-Based Polymer Materials

Maria P. Luda[1]*, *Giovanni Camino*[1], *Alexander I. Balabanovich*[1]°, *Andreas Hornung*[2]

1 Dipartimento di Chimica dell'Università Via P. Giuria 7 10125 Torino, Italy
2 Institut für Chemische Technik, Universität Karlsruhe Kaiserstrasse 12, 76131 Karlsruhe, Germany
°*Permanent address: Research Institute for Phys. Chem. Problems of Bielorussin State University ul. Leningradskaya 14, 220050 Minsk*

Summary: Pyrolytic recycling of materials for electronics and automotive is attractive because of the possibility of recovery of fuel and of precious metals from printed circuit. Due to the complexity of their composition an appropriate pre-treatment has to be performed in order to limit the evolution of dangerous halogen containing compounds which strongly impair the fuel quality. An advantageous pyrolysis approach implies the attempt of mineralisation of the organic bromine to the not volatile and harmless inorganic form using strong bases such as NaOH and KOH to reduce the amount of volatile and increasing the residue. The char stability is greatly variable depending on the substrate. $Mg(OH)_2$ and $Ca(OH)_2$ behave in a similar manner but to a lower extent. Carbonates and reducing agent such as LiAlH have been tested as well and their ability to scavenge bromine is discussed in terms of effectiveness and mechanism.

Introduction

Because of increasing use of plastics and rubber and their high visibility in wastes, polymer recycling has gained much attention in the last decades. A special branch of polymer recycling is that of mixed polymer waste. Examples for such type of waste are automotive shredder, polymer coated aluminium scrap and electronic scrap. Here the portion of polymers mingled together with other materials is often the minor part. Therefore, the polymer waste is often an undesired by-product that gives rise to appreciable problems because of the development of toxic products during the process of recycling [1]. Origin of toxics mostly comes from the use of halogenated fire retardants which were very widely used in the past because of their effectiveness. The content of halogen in forthcoming discarded computers and electronic devices is certainly decreasing, nowadays however, one have to cope with the relevant amount of fire retardant materials used in the past [2].

 CCC 1022-1360/00/$ 17.50+.50/0

There are many processes described in literature that deal with the usage of polymer fractions in various parts of foundry and cement industries and in several thermal waste treatment processes. However, it is not possible to use fraction containing electronic scrap or shredder fractions from automotive industry in the blast furnace due to the amount of halogens and metals as well as inert materials. The halogens accumulate in the cooling water systems followed by corrosion. The metals are leading to formation of alloys and the inert materials are changing the composition of the slag. The cement industries are essentially considering polymer fractions as a substitute for conventional fuel. Here, because of the alkaline conditions in the cement making process, the exhaust gases will contain only a minimum of pollutants. However, the process is limited by the amount of halogen input.

In spite of this, electronic scraps are an high value fraction due to content of noble metals and the possibility of energy recovering: even though problematic, recycling is attractive. In this contest thermal methods seem to be particularly promising. Some preliminary experimental results show that electronic scrap can be processed into halogens containing fractions, fuels with an amount of about 500 ppm halogens and residues with or nearly without halogens, depending on process parameters [3-6]. In particular, scavengers and catalysts are able to redirect pyrolysis course toward clean fuel and useful char formation. The purpose of this paper deals with the search for optimal pre-treatment of halogen scavenging.

Experimental Part

Materials

Octabromodiphenyl oxide, OBDPO, has been used as a model fire retardant. A model mixture of fire retardant high impact polystyrene, FR HIPS, has been prepared by mixing in a Brabender mixer 85% of HIPS with OBDPO (10%) and antimony trioxide Sb_2O_3 (5%). Fire retardance in this system is performed both by gas phase action [evolution of volatile $SbBr_3$ a powerful flame poison) and by condensedphase action [7-9].

Two sets of circuit board scrap, LP1 and LP2 were used. They came from computer circuit boards and from electronics for entertainment respectively. After manual separation of visible metal fragments, samples were powdered and carefully homogenised in a professional coffee mill at reduced grinding speed to avoid overheating. Selected scavengers were added to powdered samples and mixed by manual grinding in a mortar.

Techniques

The thermal decomposition was studied at a heating rate of 10°C/min under nitrogen flow by thermogravimetry using the TGA 2950 Du Pont system either in the normal mode (TG) or in the high resolution mode (HRTG; resoluton factor +5 or +6). Combination curves were calculated using Origin software.

Micro scale pyrolysis (500-1000mg of material) were performed under helium flow. The bottom part of the tube was heated up to desired temperature (heating rate 10°C/min) . Solid residues were analysed by FTIR (Perkin Elmer 2000). When appropriate, the residue was dissolved in water and filtrated. The filtrate was titrated with $AgNO_3$ 0.01N using Na_2CrO_4 as indicator.

Elemental analyses were carried out by a professional laboratory in Italy

Results and Discussion

Interaction with strong bases

OBDPO

OBDPO evaporates unaltered at 250°C when heated in quasi-isothermal HRTG mode (Figure 1, left , dashed line). In constant heating rate TG mode (10°C/min) OBDPO evaporates still unaltered between 250 and 340°C. (Figure 1, right, dashed line). When mixed with strong bases such as NaOH or KOH a residue is left which is much larger than expected on the basis of an additive behaviour. This is well demonstrated by comparing experimental and calculated thermograms. Calculated curve is a linear combination of the thermograms of the single components of the mixture, therefore it is representative of a non interacting comportment.

In order to investigate the nature of such an interaction the mixture OBDPO/NaOH 8/3 has been heated at 300°C in helium for 30 min in a microscale pyrolysis equipment. The IR of the water insoluble residue is shown in figure 2 (bottom spectrum) compared to that of original OBDPO (upper spectrum). The major difference is the appearance of a strong band at 1269 cm^{-1} which point out to the formation of Ph-O bonds.

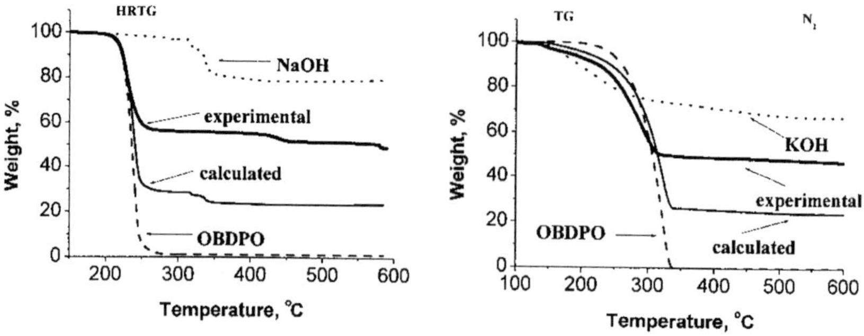

Figure 1. HRTG of OBDPO, NaOH, the mixture OBDPO/NaOH 8/3, and the correspondent calculated curve, Left; TG of OBDPO, KOH, the mixture OBDPO/KOH 16/9, and the correspondent calculated curve, right.

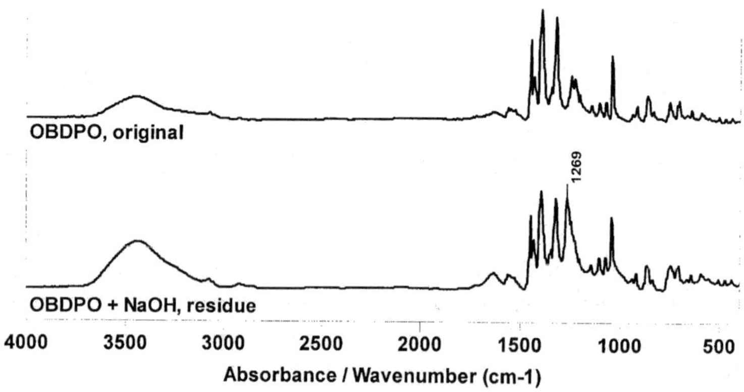

Figure 2. IR of original OBDPO (upper) and of the water insoluble residue of the thermal treatment at 300°C for 30 min of the mixture OBDPO/NaOH 8/3 (bottom).

These experimental evidences can be interpreted on the basis of the mechanism depicted in scheme 1 which shows the formation of NaBr and aromatic esters, both responsible for the residue increase. However these reactions strongly compete with volatilisation of original OBDPO.

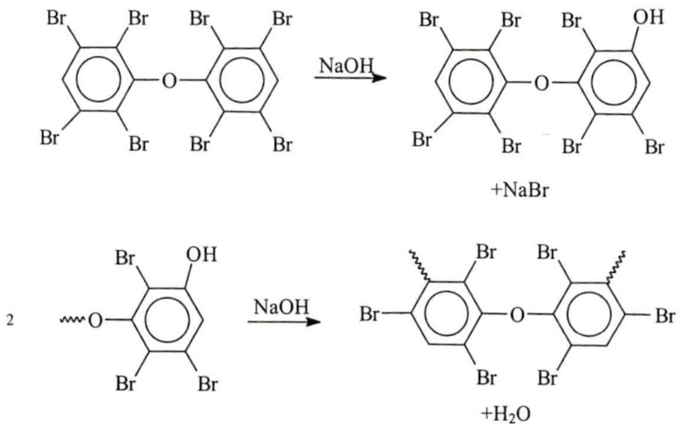

Scheme 1

FR HIPS

Interaction between NaOH and FR HIPS is well pointed out in figure 3, left. The first step of weight loss (250-350°C), which is mainly related to the evolution of volatile flame poison $SbBr_3$ [7,8] is nearly suppressed in the presence of NaOH. At the same time the residue increases. The occurrence of a preferential interaction of OBDPO with NaOH in comparison to Sb_2O_3 is strongly suggested by these data. The increase of the residue comes both by crosslinking reactions as seen for OBDPO/NaOH mixture (Scheme 1) and by avoiding antimony volatilisation.

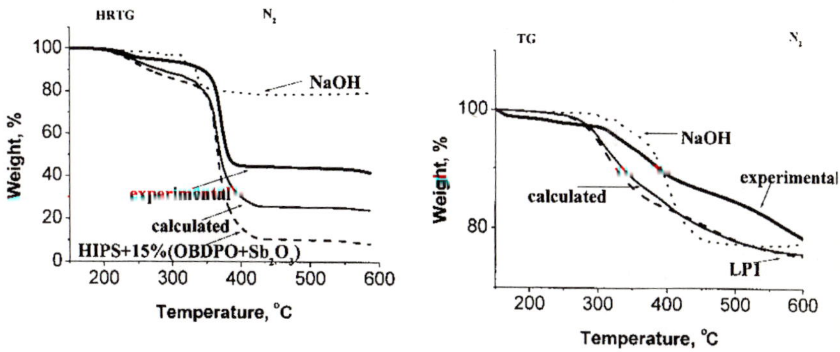

Figure 3. HRTG of FR HIPS; NaOH, the mixture FR HIPS/NaOH 78/22, and the correspondent calculated curve, left; TG LP1; NaOH, the mixture LP1/NaOH 83/17 and the correspondent calculated curve, right.

LP1

In order to clarify LP1 composition the IR on single pieces of these waste have been run. Aromatic polyester, epoxy resins polyamides and silicones have been recognised.

Their relative amount in whole LP1 waste can be highlighted from the IR of homogenised LP1 (figure 4, upper spectrum), in which aromatic polyester (1717, 1607, 1580, 1267 cm^{-1}), epoxy resins, ER, (1509, 1246, cm^{-1}) and silicon based polymers (1100, 467 cm^{-1}) can be envisaged. The spare pieces of polyamide have been also recognised prior of homogenisation which however are a minor fraction of the whole material. The complex nature of LP1 is highlighted by the continuous weight loss process.

The comparison between experimental and calculated thermogravimentric curves of figure 3, right, clearly point out at the occurrence of interaction NaOH/LP1; the weight loss is postponed to 50-100°C in the range 300-600°C. However, the residue time by time formed is not thermally stable. In order to investigate the nature of such an interaction LP1 has been treated 2 hours with NaOH at 300°C. Results on parallel heating on LP1 and on LP1/NaOH mixture are reported in table 1.

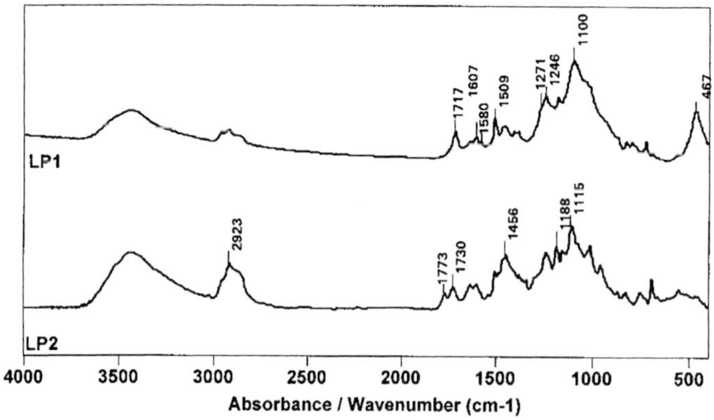

Figure 4. IR of homogenised powdered mixtures LP1 (upper) and LP2(bottom).

Table 1. Residue composition from thermolysis at 300°C for 2 hours of LP1 and LP2 in different experimental conditions.

	LP1	LP1 +NaOH 17%	LP2	LP2 +NaOH 17%
Cl initial	0.14%	0.12%	2.30%	1.91%
Br initial	3.7%[1]	3.2%[1]	2.22%	1.84%
Halogen, initial meq/g	0.5	0.41	0.94	0.78
Residue amount	82%	96%	77%	88%
Water soluble residue	1.4%	27%	1.1%	27%
Water soluble halogen	27%[2]	80.5%[2]	14.8%[2]	62.1%[2]

1: elemental analysis; % refers to the initial amount in the mixture
2 $AgNO_3$ titration; % refers to the initial amount of halogens(meq) in the mixture

The residue (96%) of the mixture is partially water soluble (27% of the initial sample, table 1). Titration with $AgNO_3$ of the soluble residue shows that 80.5% of the initial equivalents of halogen are converted to inorganic soluble salts. The IR of the residue (figure 5, upper spectrum) is dominated by the absorption of the carboxylate (1574, 1440 cm^{-1}), carbonate (1440 and 880 cm^{-1}) and possibly bicarbonate (1278 cm^{-1}) ions. All these absorption disappears as soon as the residue is washed with water (figure 5, spectrum in the middle). The structure of this residue shows similar features with that obtained by heating LP1 at 340°C without NaOH (Figure 5, bottom spectrum). However a new absorption occurs at 1037 cm^{-1}, which can be attributed to the ether groups formation.

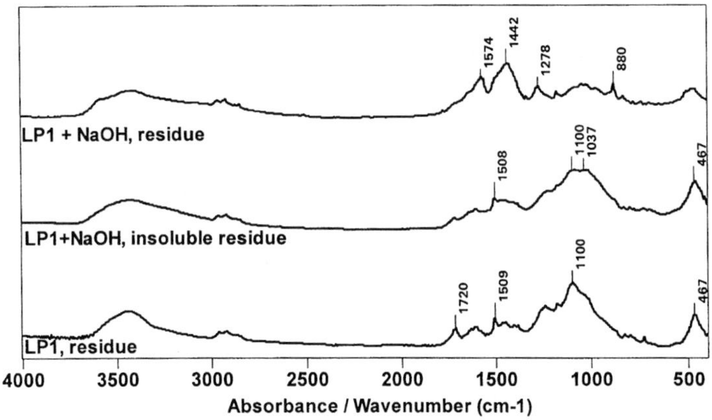

Figure 5. IR spectrum of the residue after heating in helium at 300°C for 2 hours the mixture LP1/NaOH (83/17), up; of the water insoluble residue, middle; and of the residue of LP1 after heating in nitrogen up to 340°C at 10°C/min , bottom.

A possible mechanism is proposed in scheme 2.

Organic carboxylates and aliphatic alcohols are formed by basic catalysed hydrolysis of polyesters. decarboxylation can also occur by heating (scheme 2, reactions 1). In such a strong basic entourage dehalogenation of aromatic structures takes places (scheme 2, reaction 2). Phenols sodium salts interact with water and CO_2 forming sodium bicarbonate which in turn can be converted to carbonate by heating. Eventually condensation of phenate and ROH leads to the ether formation shown by IR (reaction 3).

The amount of soluble residue depends on the balance between soluble carboxylates and NaBr formation, residual NaOH and condensation reactions.

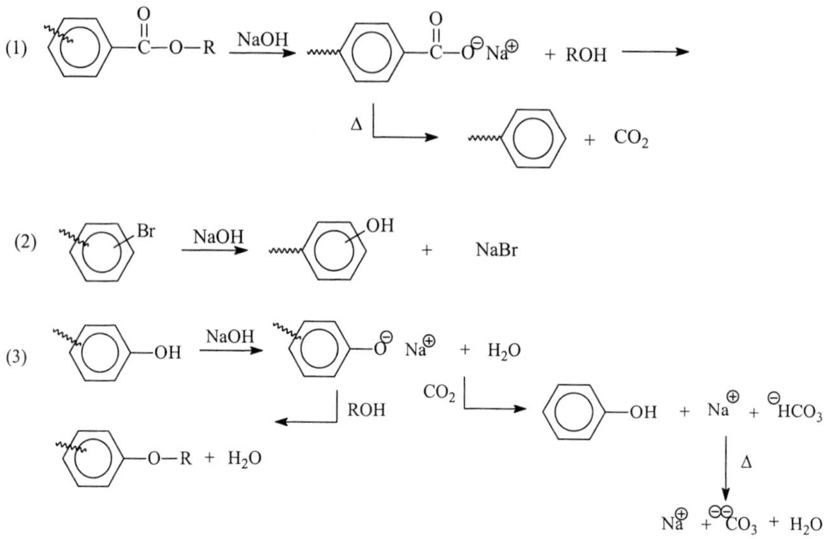

Scheme 2

LP2

The main constituents of LP2 are polyolefines (2923, 1456 cm^{-1}), polyphenylenoxide, PPO, (1188 cm^{-1}), polycarbonates, PC, (1773 cm^{-1}) and polyesters (1730 cm^{-1}) possibly containing polyols (1115 cm^{-1}) (Figure 4, bottom spectum). Acrylonitrile based polymers (2238 cm^{-1}) have also been detected in spares pieces of LP2 as a minor fraction.

The same qualitative features described so far for LP1/NaOH interaction have been found for thermal treatment of LP2/NaOH mixture (table 1): the residue of LP2/NaOH mixture is partially water soluble and 62% of the initial amount of halogen (in equivalent) is transformed in water soluble salts.

Interaction with $Ca(OH)_2$ and $Mg(OH)_2$

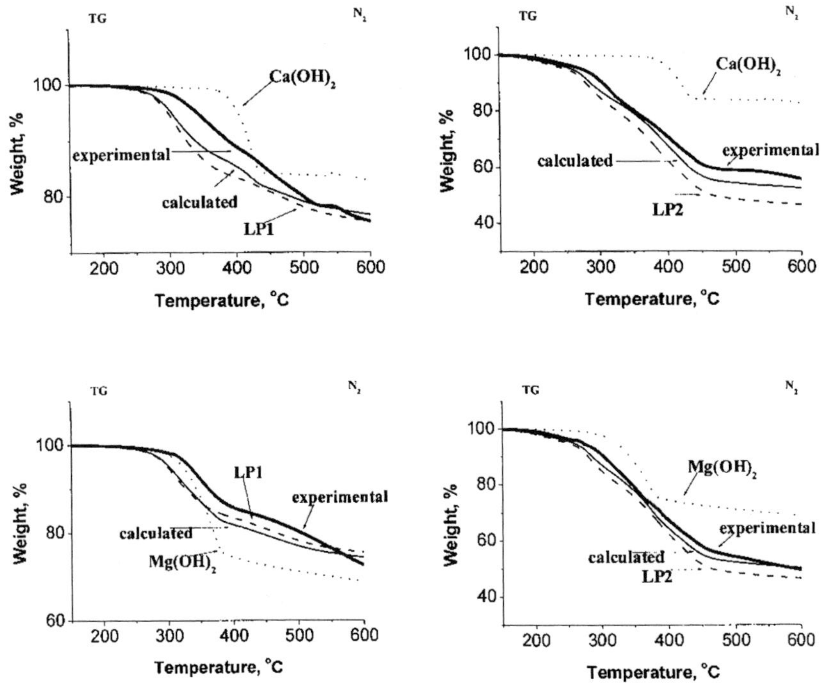

Figure 6. TG of LP1 and LP2 in mixture with 17% of $Ca(OH)_2$, first row, and with 17% of $Mg(OH)_2$, second row.

$Ca(OH)_2$ and $Mg(OH)_2$ are weaker bases than NaOH and their ability as halogen scavenger is greatly reduced. This is in agreement with the nucloephylic character of aromatic substitution in scheme 1 and 2 which require strong nucleophylic agents and activated substrates. They are ineffective with OBDPO and very poorly effective with LP2 in which calculated and experimental thermograms nearly overlap (Fig. 6, right column). In LP1, degradation is postponed to 20-40°C in the presence of these bases but the residue is not stable at the temperature at which is formed and eventually a smaller residue is left (Fig. 6 left column). Thus, $Ca(OH)_2$ and $Mg(OH)_2$ are able to interact only with more reactive species which comes from degradation of LP1 (eg. HBr or aliphatic bromine containing products) (10) and therefore might be reactive in aggregation with agents catalysing HBr formation. Investigation on the volatile degradation products is currently in progress in order to elucidate this point.

Interaction with $CaCO_3$

$CaCO_3$ interact very poorly with OBDPO and LP2 and better with LP1 (figure 7, left) and FR HIPS. The same conclusion discussed so far for weaker bases $Ca(OH)_2$ and $Mg(OH)_2$ can be drawn in this case. For example it is clearly imaginable that, if HBr is evolved during thermal degradation, it will react immediately with $CaCO_3$ forming $CaBr_2$.

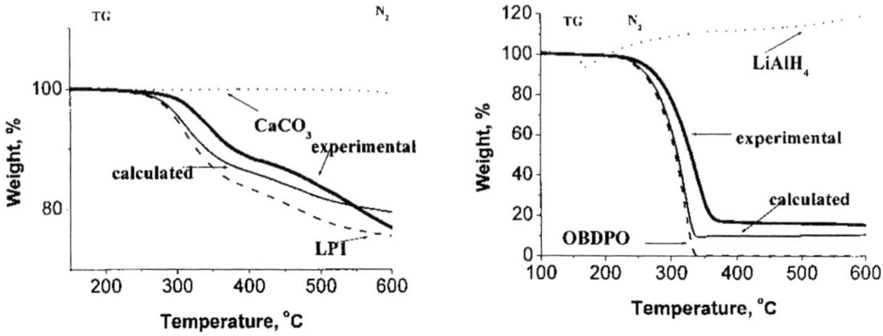

Figure 7. Thermograms of LP1 in mixture with 17% of $CaCO_3$. (left) and of OBDPO in mixture with 9% of $LiAlH_4$.

Interaction with reducing agents

TG of stoichiometric mixture of $LiAlH_4$/OBDPO shows that interaction between them occurs (Figure 7, right): the residue left at 350°C fits with the stoichimetry reported by Brown at al [11]

$$Ar\text{-}Br + 2\ LiAlH_4 \rightarrow 2\ Ar\text{-}H\ (volatile) + LiBr\ (unvolatile) + Li\ Al_2H_7\ (unvolatile)$$

Evaluation of the degradation products will help in clarifying whether this mechanism really occur or not. If this is the case, this reaction will produce very clean fuel whereas the residue is fully inorganic. However the difficulties arising from $LiAlH_4$ manipulation in industrial environment and costs will prevent from using this agent as a scavenger despite its possible large effectiveness.

Conclusions

Compounds of different chemical classes have been tested as halogen scavenger for clean pyrolysis of real electronic scraps and model mixtures.

Strong bases such as NaOH and KOH are highly effective by leaving at a reasonably low temperature a large residue containing mineral halogens. Therefore, this residue is suitable for a second-step-pyrolysis at higher temperatures to get a large amount of clean fuel. However, strong bases can be aggressive for the recycling implants.

Weaker bases and $CaCO_3$ are able to capture only reactive species evolved during degradation such as HBr or aliphatic halides and, therefore, can only be useful for scavenging halogen from reactive substrates; reducing agent can be very effective in principle but costs and environmental consideration prevent their large employ.

Investigations are in progress to fully define the interaction mechanism by analysis of volatile products of thermal treatment and to find less aggressive substitute for strong bases. In this view, a mixture $Na_2CO_3/Ca(OH)_2$ which should form *in situ* NaOH is actually under investigation.

Acknowledgement

The authors are grateful to the European Community for funding this work in the frame of competitive and sustainable growth (Growth) programme G1RD-CT-1999-00082.

[1] A.G. Buekens, *Proc. UNIDO Conference*, Copenhagen **1991** US/CPR/86/130.

[2] H. Buser, *Environ. Sci Technol.* **1986**, *20(4)*, 404.

[3] H. Bockhorn, A. Hornung, U. Hornung, *J. Anal. Appl. Pyrolysis*. **1999**, *50*, 77.

[4] H. Bockhorn, A. Hornung, U. hornung, P. Jakobströer M. Kraus, *J. Anal. Appl. Pyrolysis* **1999**, *49*, 97.

[5] H. Bockhorn, A. Hornung, U. Hornung, *J. Anal. Appl. Pyrolysis*. 1998, *46*, 1.

[6] H. Bockhorn, A. Hornung, U. Hornung, P. Jakobströer, *Combust. Sci. and Tech* **1998,** *134,* 7.

[7] L. Costa, G. Camino, M.P. Luda, *Polym. Deg. Stab.* **1986**, *14,* 113.

[8] M.P. Luda, G. Camino, L. Costa, P. Roma, A. Rossi, *J. Anal. Appl. Pyrolysis*. **1987**, *11*,511.

[9] L. Costa, M.P. Luda and L. Trossarelli *Polym. Deg. Stab.* **2000**, *68*,67.

[10] M.P. Luda, A.I. Balabanovich, G. Camino, *J. Anal: Appl. Pyrolysis*. Submitted.

[11] H.C. Brown, J. Khrishnaumurthy. *Org. Chemistry* **1969,** *34,* 3919.

Macromol. Symp. ***180,*** *153–167 (2002)*

Self-Toughening Liquid Crystalline Vinyl Ester Adhesives

E. Amendola[1,*], *M. Giamberini*[1], *C. Carfagna*[2], *V. Ambrogi*[2]

[1] Institute of Composite Materials Technology, CNR.
P.le Tecchio 80, 80125 Naples - ITALY

[2] Dept. of Materials and Production Engineering, University of Naples "Federico II".
P.le Tecchio 80, 80125 Naples - ITALY

Summary: Vinyl ester resins are widely used as thermosetting matrices for production of advanced composite and structural adhesives. They are obtained by end-capping unsaturated vinyl groups on a large variety of suitable oligomeric precursor, resulting in thermosetting resin with a good processability and outstanding end-use properties. In this paper we describe the shynthesis of a new methacrylated derivative of p-(2,3-epoxypropoxy)-α-methylstilbene exhibiting liquid crystalline phases, referred to as MD3A2. For the sake of comparison, a model compound has been synthesized with very similar chemical backbone. The model compound, referred to as MEP3A2, features isotropic phase and is derived form the commercially available Epon 825® epoxy resin (Shell Oil Company). Fracture resistance of samples has been tested according to ASTM D1002-72 (Strength Properties of Adhesives in Shear by Tension Loading) on aluminum supports and related to the presence of isotropic or liquid crystalline phases.

Keywords

Structural adhesive, liquid crystalline thermoset, vinyl ester resin, fracture resistance.

Introduction

Vinyl esters are thermosetting resins prepared by reacting acrylic or methacrylic acid with a suitable oligomeric precursors, including urethane, novolac, bisphenol and alogenated bisphenol resins [1-3]. They exhibit end-use properties and raw material costs which are intermediate between the high performance epoxy and the cheap unsaturated polyester resins (UPE).

The end-use properties of vinyl ester resins are strongly influenced by the chemical nature of starting precursors, while the presence of unsaturated vinyl end-groups affects the crosslinking reaction and the processability of the resin.

 CCC 1022-1360/00/$ 17.50+.50/0

The most widely used vinyl esters are the epoxy based ones, which find useful application as matrices for advanced composites [4-10].

Introducing unsaturated terminations, the curing process changes from the step reaction mechanism to the free radical polymerization, similar to that occurring with unsaturated polyesters. In the case of radical crosslinking reaction, gelation occurs at a much lower fractional conversion with respect to step reaction mechanism (i.e. epoxy resins). This leads to sensible processing advantages, enabling the production of large manufacts cured at room temperature. It has been found that vinyl ester resins cured at room temperature can achieve up to 80-90% of their optimal mechanical properties. Moreover, the epoxy backbone confers better toughness, elongation and chemical resistance to vinyl ester resins compared to unsaturated polyesters. These features are mostly due to the presence of ether groups along the molecular chain, unlike polyesters in which ester linkages are present. Ester groups are more sensitive to hydrolysis than ether groups, especially in the presence of strong bases and acids. In vinyl ester resins the ester linkages are produced by the addition reaction of unsaturated acids to epoxy end-groups. In the case of methacrylated resins, the steric hindrance of large methyl groups next to each ester linkage largely improves the hydrolytic resistance of material.

The adhesion of vinyl esters to solid substrates strongly benefits from the presence of secondary –OH groups in β–position to the C=O groups, resulting from the addition reaction of the carboxylic acid to the epoxy ring. As a result, vinyl ester resins wet effectively and strongly bond to solid substrates and reinforcing fibers. Long-term durability, resistance to thermal, oxidative and hydrolytic degradations, and bond retention after extended exposure to moisture are essential requirements of both structural adhesives (which must be highly crosslinked, hence harder and more brittle) and protective coatings (for which moisture impermeability and corrosion protection are required).

Processability of vinyl ester resins can be further improved by the addition of a co-reactive solvent, usually styrene, which drops the viscosity down to 100-500 mPa*s at room temperatures, thus allowing a rapid and effective wet-out of the glass fiber. The co-reactive solvent is incorporated into the polymeric backbone during crosslinking reaction. Depending on their specific properties [11-13], which are related to solvent content, the epoxide backbone, and the kind of termination, vinyl esters find useful applications as glass fiber composites, adhesives, bulk molding and sheet molding compounds for appliance housings and automotive components. In particular, due to the high reactivity of the double bonds, acrylated vinyl ester resins are widely used for UV curable inks and coatings.

In this paper, the synthesis and properties of a liquid crystalline vinyl ester resin are presented and compared to a suitable isotropic model compound. The liquid crystalline and the isotropic resins have been tested as adhesive for aluminum substrates, in order to evaluate their use as adhesives for strong metal-to-metal structural joints and durable protective coatings. In fact, structural adhesives offer a significant alternative to conventional metal joining techniques, with lower prices, higher durability, faster application and better resistance.

Recently, it has been shown that mechanical properties of epoxy resins can be strongly improved if a liquid crystalline structure is induced to the network [14-23]. The resulting materials are characterized by high toughness, high chemical resistance, high resistance to water permeability and, mainly, by the possibility of being oriented by magnetic or electric fields before or during curing [24-25]. This class of materials offers significant improvements over conventional resins for advanced applications.

Experimental Part

Synthesis

The p-(2,3-epoxypropoxy)-α-methylstilbene (DOMS) was synthesized according to procedures reported in the literature [26]. The synthesis of MD3A2 was carried out starting from the epoxy monomer DOMS, whose liquid crystalline properties in thermosets have been extensively studied and previously reported [27-31].

The chemical structure of MD3A2 is shown in figure 1.

MEP3A2

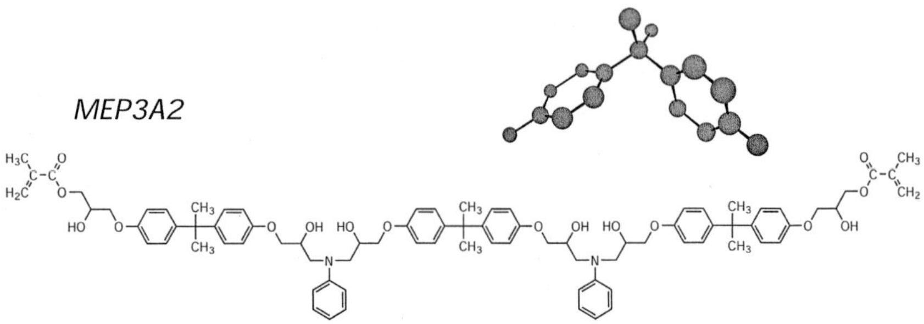

MD3A2

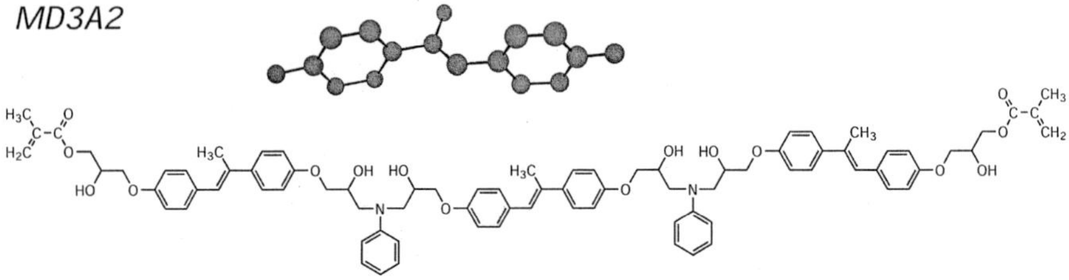

Fig. 1. Structural formulae of MD3A2 and MEP3A2; pictorial view of molecular moieties

In a typical procedure, DOMS (10 grams) was reacted with aniline (1.83 g) in a molar ratio of 3 to 2, using glacial acetic acid (5-7 drops) as catalyst and dioxane (20 ml) as solvent, at reflux temperature for 3 hours. In the second reaction step, 1.86 g of methacrylic acid (10% molar excess) were added, together with 1 mg of hydroquinone, to prevent the polymerization of methacrylic acid, and a few drops of N,N-benzyldimethylamine, as catalyst. The progress of the reaction between DOMS and methacrylic acid was followed by FT-IR. One droplet of reaction mixture was collected on KBr crystals every 30 minutes and analyzed, monitoring the decrease of the peak at 914 cm^{-1}, characteristic of the epoxy group. After 2 hours the product was precipitated into 1 liter of cold hexane. It was then filtered and washed with isopropanol (200 ml), filtered and washed with hexane (300 ml), filtered and vacuum dried at room temperature.

The final product was a white solid.

Yield: 67%. T_m=45°C (ΔH=29 J/g) T_i=94°C (ΔH=1.6 J/g). The melting and the isotropization temperatures, indicated with T_m and T_i respectively, are reported as the onsets of the calorimetric trace.

MEP3A2 is the methacrylic derivative of the epoxy monomer, synthesized from a commercial epoxy resin, Epon 825® and aniline in a molar ratio of 3 to 2, according to the same procedure followed for MD3A2.

Aniline was distilled before its use. All chemicals were purchased from Aldrich.

Sample preparation

MD3A2 and MEP3A2 were cured in the presence of benzoyl peroxide (1% wt.) as initiator and cobalt naphthenate (0.1% wt.) as catalyst at 150°C for 4 hours.

Aluminum substrates were pretreated according to the ASTM standard 2651-90 (Standard Guide for Preparation of Metal Surfaces for Adhesive Bonding).

Degreasing was accomplished by wiping with a cloth dipped in 1,1,1,trichloethane. In addition to solvent degreasing, a chromic acid etching technique was carried out on the substrates. A solution of distilled water, sulfuric acid, and sodium dichromate dihydrate in a weight ratio of 100:30:6, respectively was used, commonly referred to as FPL solution. After degreasing and drying, the substrates were immersed in the FPL solution at T=70°C for 12 minutes, washed with cold running tap water and dried at 60°C for 30 minutes.

Characterization

Seiko differential scanning calorimeter (DSC) mod. 220C, was used to evaluate the melting and the isotropization temperatures, and the heat flows involved in the transition of monomers. The experiments were run at 10°C/min heating rate under nitrogen flow.

Morphological features of starting monomers and cured resins were observed in transmitted light with an optical microscope Reichert-Jung (mod. Polyvar) equipped with crossed polarizers. The temperature was controlled by means of a Linkam mod. TH 600 hot stage.

X-Ray diffraction patterns were recorded with a flat camera equipped with a modified Linkam THMS 600 hot stage, with a sample to camera distance of 8.0 cm (Ni-filtered Cu-Kα radiation). The Fujifilm MS 2025 imaging plate and a Fuji Bio-imaging Analyzer System, mod. BAS-1800, were used for digitizing the diffraction patterns.

Fourier Transform Infrared spectrometer (FTIR) Nicolet mod. 5PC at a resolution of 2 cm^{-1} was used for spectroscopic analysis.

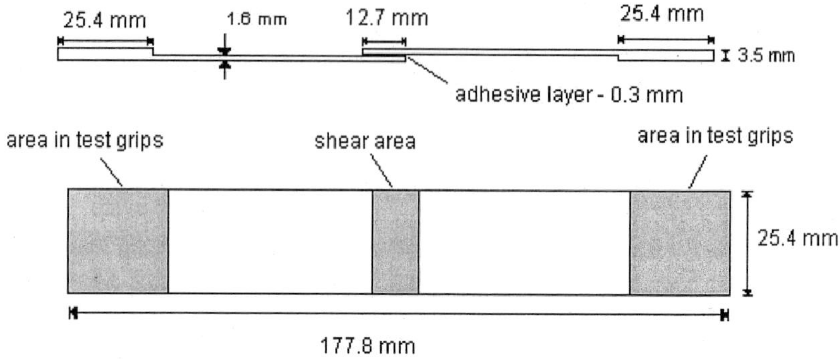

Fig. 2. Dimensions of substrates according to ASTM D1002-72

Strength of adhesive bonding was tested according to ASTM standard D 1002-72 (Strength Properties of Adhesives in Shear by Tension Loading). The shape and dimensions of the test specimens are shown in figure 2.

An electro-mechanical dynamometer INSTRON 4204 was used for mechanical testing, with a 5 kN load cell. Deformation rate was 1.3 mm/min. The specimens were fitted in such a way that their long axis coincided with the direction of applied deformation.

Dynamical Mechanical characterization has been carried out by means of TA Instruments DMA 2980, at a frequency of 1 Hz in tensile mode. Glass transition has been associated to loss modulus maximum.

Results and Discussion

Thermal analysis has been performed on MD3A2 by DSC. The first heating scan shows a solid-to-liquid crystalline phase transition centered at T=55°C (Tonset=45°C, ΔH=29 J/g) and a liquid crystalline-to-isotropic phase transition at T=101°C (Tonset=94°C, ΔH=1.6 J/g). The X-rays diffraction pattern obtained on the sample at T=75°C and the analysis of textures with optical microscope indicate the formation of a nematic phase in the temperature range between the melting and the clearing temperatures. The crosslinking reaction takes place after isotropization, as indicated by the exothermic peak centered at T= 140°C (Tonset=124°C, ΔH=74 J/g).

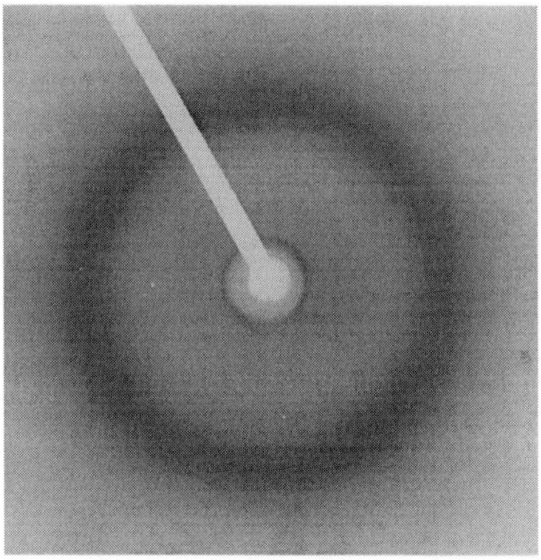

Fig. 3. X-ray diffraction pattern of cured MD3A2 recorded at room temperature

In the second heating scan the curve shows a glass transition at T=60°C and an endothermic peak centered at T=137°C (Tonset=123°C, ΔH=2.7 J/g), this latter corresponding to the isotropization of the resin. Upon curing, a LC polymeric network forms in which chains are still free of moving giving rise to the reversible LC-to-isotropic transition above T=123°C. The X-ray pattern obtained on the cured sample at room temperature indicates the formation of a smectic structure, with a sharp reflection at low angle corresponding to a spacing d=20.5Å, which is typical of a smectic phase (see figure 3).

MD3A2 has been used as adhesive for aluminum joints. Its features have been compared to those of MEP3A2, chosen for the comparison. The chemical structures of the two monomers are very similar. However, due to the presence of an unsatured group between the two aromatic rings, MD3A2 exhibits a rigid structure, which is responsible for the formation of the LC phase. Differently, MEP3A2 has a more flexible chain, and thus exhibits isotropic properties (see figure 1).

To study the adhesion properties on aluminum joints of liquid crystalline resin in comparison with an isotropic one, a three-step procedure, consisting in the pre-treatment of aluminum supports, application of the resin, mechanical characterization of joints has been carried out.

Like any other metal, aluminum exhibits a very high surface energy (E>500 mJ/m2) and hence it reacts easily with the oxygen in the atmosphere giving rise to the formation of the corresponding oxide [32-34]. The oxide layer formed can isolate the underneath metal layer, thus reducing the interaction between the metal and the polymer used as adhesive. Furthermore, the oxygen atoms on the surface can easily hydrate in the presence of humidity in the atmosphere, according to the reaction (1):

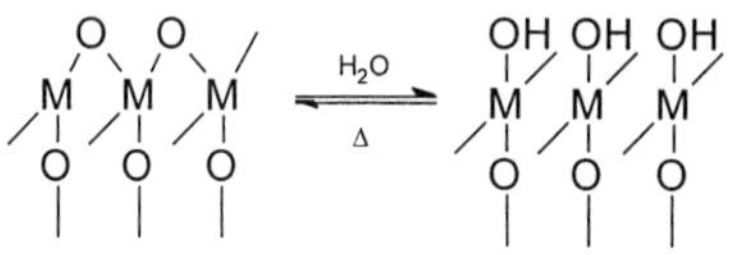

(1)

The presence of hydroxyl groups can damage the joint by favoring the adsorption of water and polar contaminant. In particular, water can be strongly detrimental because hydrated aluminum oxides adhere weakly to the metal.

A solvent degreasing pre-treatment is required to remove surface contamination, such as machine oils, grease, release agents. It is usually followed by a strong acid etching. This latter provides a very rough metallic surface which favors the *mechanical interlocking* between the resin and the metallic substrate, and in the case of resin with quite low viscosity, a good penetration of the adhesive inside the small cavities and inhomogeneities formed [35]. The procedure followed for pre-treatment of joints is explained on the ASTM standard D2651-90 (Standard Guide for Preparation of Metal Surfaces for Adhesive Bonding). Efficient removal of weak boundary layers, improved interfacial contact (better wetting and a greater area for interfacial contact) and the enhancement of energy dissipative mechanisms in the adhesive are obtained [36]. All these mechanisms contribute to improve the resin/metal adhesion and thus to increase the joint strength. To prevent the formation of new detrimental hydrated layers, the resin had to be applied as soon as the substrates were ready.

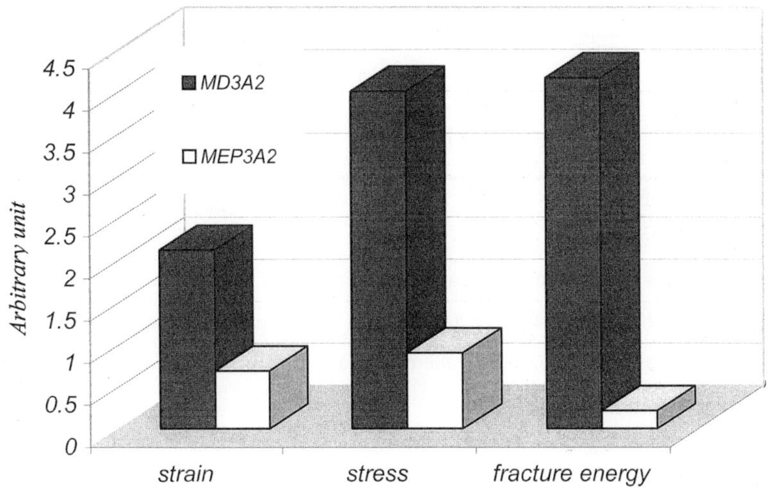

Fig. 4. Comparison between mechanical properties of MD3A2 and MEP3A2

After the pre-treatment, the resin has been readily applied on supports. Before being applied, the resin has been mixed with the benzoyl peroxide (1% wt.) used as radical initiator, and the cobalt naphtenate (0.1 % wt.) used as catalyst. The resin between joints was cured in oven at T=150°C for 4 hours.

The adhesion properties of joints have been tested according to the ASTM D1002-72 (Strength Properties of Adhesives in Shear by Tension Loading).

The characterization of liquid crystalline and isotropic resins is summarized in table 1. The glass transition temperature and elastic modulus of MD3A2 have been calculated by DMA analysis. The elastic modulus value, calculated at 25°C, is in good agreement with the expected value for and amorphous glassy polymer. Even though the sample exhibits a liquid crystalline phase, the random domain orientation does not contribute to the modulus increase in the elongation direction. The value of single lap shear strength is in agreement with literature data.[37]

Table1. Physical chemical characterization and mechanical properties of MD3A2 and MEP3A2.

Nature of resin	sample	T_g (°C)	E (GPa)	Max. stress (MPa)	Max. strain	Energy
LC	**MD3A2**	62 *	2.5 *	4.0	2.7	19
ISO	**MEP3A2**	47 *	1.2 *	0.9	unity	unity

* result of DMA characerization.

The last two columns in table 1 report mechanical testing results. Due to the presence of defects in the resin bulk, and the absence of a suitable strain gauge for the accurate measure of sample deformation, the results are expressed as the ratio of the actual values of LC resin versus isotropic one.

Figure 4 shows the comparison between mechanical properties of MD3A2 and MEP3A2 used as adhesives.

A big contribution to the interpretation of mechanical test results is provided by SEM images of fractured surfaces. Figure 5 shows the fracture surface of MD3A2, which appears rough, with a microfibrillar structure. It is possible to hypothesize that the fracture mechanism is related to the peculiar structure of liquid crystalline thermosetting resins. The microstructure of LC thermosets is characterized by locally ordered regions, so that when the material starts deforming the crack is forced to deviate continuously along the differently oriented domains. This kind of fracture has been previously observed on other liquid crystalline thermosetting resins and composites [38-40]. The mechanism is comparable to that of rubber modified thermosets, which is typical ductile fracture.

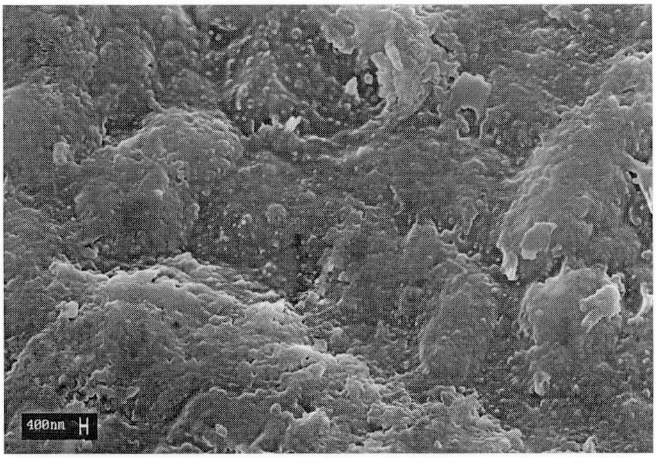

Fig. 5. SEM micrograph of fratcurated MD3A2

This feature can be explained taking into account the particular microstructure of cured liquid crystalline thermosets (LCTs) [41, 42]. An LC network is characterized by a polydomain morphology, in which portions of liquid crystalline domains are embedded in an isotropic matrix. LCTs tend to form a macroscopically disordered polydomain in the absence of external fields. The presence of such domains causes a ductile fracture: this inhomogeneity is responsible for the outstanding toughness of this class of materials.

This result was also found by Sue et al. who studied the fracture behavior of a liquid crystalline resin based on the diglycidyl ether of 4,4'-dihydroxy-a-methylstilbene cured with sulfanilamide [42]. In their paper, the authors observed crack segmentation, crack branching, crack bridging, and crack deflection.

The behavior of MEP3A2 is different, as confirmed by SEM image (figure 6). The fracture surface is smooth, which is typical of fragile materials: the crack propagates quickly and linearly, leaving the surface homogeneous and smooth.

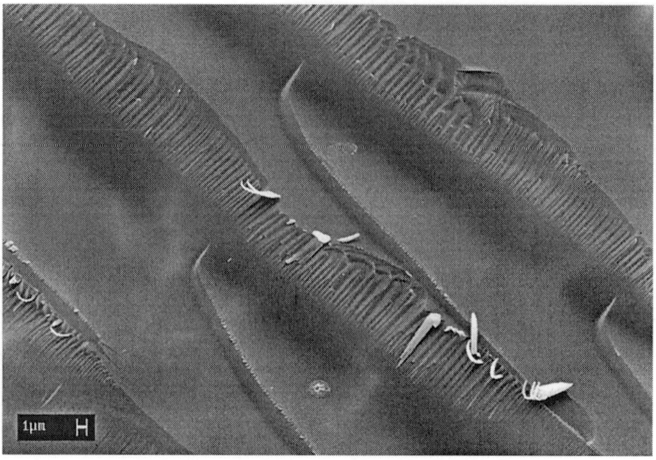

Fig. 6. SEM micrograph of fratcurated MEP3A2

From SEM analysis it is also evident that both in MD3A2 and MEP3A2 case, a cohesive rupture occurs, which is located in the bulk of the resin and not at the interface. This is related to several factors, the most significant of them are:

- Non-bonding interactions between the resin and metal oxides:

The adhesion between the resin and the substrate is mostly due to the presence of hydrogen bonds which form between the oxide layer and both the hydroxyl groups in β-position to the carboxylic group, and ether groups.

Furthermore, in a work carried out by Holubka et al. [44], according to theoretical studies based on semiempirical computational models, it has been shown that methacrylic monomers, as well as acrylic ones, can interact with aluminum oxides, both through the carbon-carbon double bond C=C, and the carbon-oxygen double bond C=O (figure 7). The basic idea is that the aluminum oxide and the unsatured material act as Lewis acid and Lewis base respectively. The reaction occurs through the formation of intermediates, which favor the adhesion of the resin to the oxide immediately before the curing reaction. The stability of these structures can affect the final features of adhesives.

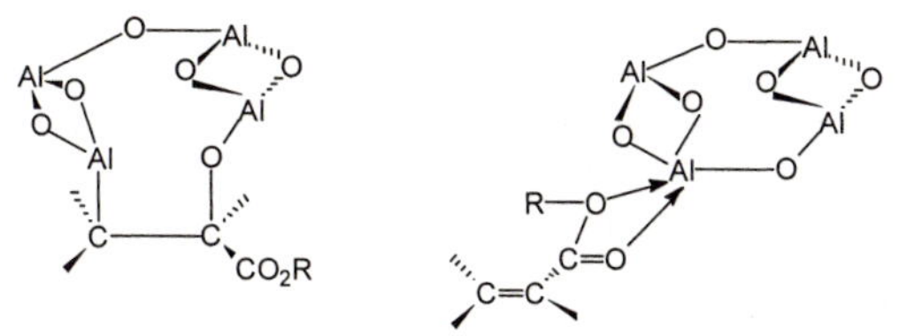

Fig. 7. Schematic representation of interactions between resin and aluminum oxide

- Pre-treatment of metallic substrates:

Another factor is related to the treatment performed on the metal substrate before the joint is prepared. The surface must be as rough and porous as possible to insure a good adhesion with the resin. When the resin melts during the cure, it can easily penetrate the cavities formed, so that when the adhesive is completely cured, it turns out to be embedded in the oxide structure. An interfacial layer with composite structure forms, that can better dissipate the energy when is deformed, thus improving the adhesion strength.

Conclusions

In the last decades, adhesive bonding has achieved rapid growth, due to the numerous advantages that structural adhesives offer over conventional joining methods.

In this work, it has been shown that high adhesion strength can be achieved by using liquid crystalline adhesives. The maximum strain and stress obtained was much higher for the joints with the liquid crystalline adhesive than with the isotropic one.

The different fracture behaviors have been related to the microstructures of the two cured resins.

The mechanical interlocking and the molecular interactions between the resin and the metallic substrates have given rise to the formation of a very strong and stable interface layer in which the adhesion forces are stronger compared with the internal cohesive forces in the bulk of the resin, so that the rupture under mechanical stress occurred within the resin and not at the interface with the metallic substrate, producing a cohesive rupture.

Moreover, due to the higher toughness observed in the case of liquid crystalline resin, the adhesion strength of this latter turns out to be higher compared to that of the isotropic resin.

References

[1] M.B. Launikitis, in: "*Handbook of Composites*". Van Nostrand Reinhold, New York (1982), p 38.
[2] P. Kelly, in: "*Reinforced Plastics Durability*". CRC Press, Boca Raton, FL (1999), p 282.
[3] S.H. Goonmann, in: "*Handbook of Thermoset Plastics*". Noyes Publications, Park Ridge, NJ (1986).
[4] R.D. Patel, J.R. Thakkar, R.G. Patel and V.S. Patel, *High Performance Polymers* 2, (1990), p 261
[5] B. Gaur and J.S.P. Rai, *Polymer* 33, (1992), p 4210.
[6] P. Siva, I.K. Varma, D.M. Patel and T.J.M. Shina, *Bull. Mater. Sci.* 17, (1994), p 1095.
[7] X. Dirand, B. Hilaire, E. La Fontaine, B. Mortaigne and M. Nardin, *Composites* 25, (1994), p 645.
[8] J.S. Ullett and R.P. Chartoff, *Polym. Eng. Sci.* 35, (1995), p 1086.
[9] S. Pham and P.J. Burchill, *Polymer* 36, (1995), p 3279.
[10] R.R. Pachha, J.R. Thakkar and R.D. Patel, *High Perform. Polym.* 5, (1993), p 207.
[11] R. Bhatnagar and I.K. Varma, *J. Therm. Anal.* 35, , (1989), p 1241.
[12] T.C. Goel, A. Tripathi, B.S. Rao, M.S. Choudhary, V. Choudary and I.K. Varma, *J. Appl. Pol. Sci.*, 30, (1985), p 1491.
[13] Kant, A. Mishra and J.S.P. Rai, *Polym. Int.*, 28, (1992), p 189.
[14] Hefner, JR., J.D. Earls and P.M. Puckett, *US Patent 4,962,163* (1990).
[15] Dhein, H.P. Muller, H.M. Meier and R. Gipp, *US Patent 4,762,901* (1988).
[16] Jones, JR., J.D. Earls and R.E. Hefner, Jr., *US Patent 5,248,360* (1993).
[17] Guymon, R. Shao, D. Holter, H. Frey, N.A. Clark and C.N. Bowman, *Liq. Cryst.* 24, (1998), p 263.
[18] R.A.M. Hikmet, B.H. Zwerver and D.J. Broer, *Polymer* 33, (1992), p 89.
[19] R.A.M. Hikmet and D.J. Broer, *Polymer* 32, (1991), p 1627.
[20] R.A.M. Hikmet, *Liq. Cryst.* 9, (1991), p 405.
[21] R.A.M. Hikmet, *Macromolecules* 25, (1992), p 5759.
[22] S.M. Kelly, *Liq. Cryst.* 24, (1998), p 71.
[23] C.E. Hoyle, T. Watanabe, and J.B. Whitehead, *Macromolecules* 27, (1994), p 6581.
[24] E.P. Douglas, D.A. Langlois and B. Benicewicz, *Chem. Mater.* 6, (1994), p 1925.
[25] D.A. Langlois, B. Benicewicz and E.P. Douglas, *Chem. Mater.* 10, (1998), p 3393.
[26] M. Giamberini, E. Amendola and C. Carfagna, *Mol. Cryst. Liq. Cryst*, 266, (1995), p 9.
[27] C. Carfagna, E. Amendola, M. Giamberini, A.G. Filippov and R.S. Bauer, *Liq. Cryst.* 13, (1993), p 571.
[28] E. Amendola, C. Carfagna, , M. Giamberini and G. Pisaniello, *Macromol. Chem. Phys.* 196, (1995), p 1577.
[29] G. Micco, M. Giamberini, E. Amendola, C. Carfagna and G. Astarita, *Ind. Eng. Chem. Res.* 36, (1997), p 2976.
[30] M. Giamberini, E. Amendola and C. Carfagna, *Macromol. Rapid Commun.* 16, (1995), p 97.
[31] B.C. Benicewicz, M.E. Smith, J.D. Earls, R.D. Priester, JR., S.M. Setz, R.S. Duran and E.P. Douglas, *Macromolecules*, 31, (1998), p 4730.
[32] A.J. Kinloch, in: "*Structural Adhesives: Developments in Resins and Primers*". Elsevier, New York (1986).
[33] J.S. Ahearn, G.D. Davis, T.S. Sun and J.D. Venables, in: "*Adhesion Aspects of Polymeric Coatings*". Plenum Press, New York (1983), p 281.
[34] J.S. Noland, *Adhesion Science and Technology*, Part A, Plenum Press, New York (1975), p. 413.
[35] B.W. Malpass, D.E. Packham and K. Bright, *J. Appl. Polym. Sci.* 18, (1974), p 3249.
[36] A.J. Kinloch, *Adhesion and Adhesives: Science and Technology*, Chapman and Hall, New York (1987).

[37] N. Cadinot, B. Boutevin, J.P. Parisi, D. Béziers, E. Chataignier, *Int. J. Adhesion and Adhesives,* 14 (4), 237 (1994).

[38] C. Carfagna, E. Amendola and M. Giamberini, *Prog. Polym. Sci.* 22, (1997), p 1607.

[39] C. Carfagna, E. Amendola, M. Giamberini, A. D'Amore, A. Priola and G. Malucelli, *Macromol. Symp.* 148, (1999), p 197.

[40] C. Carfagna, E. Amendola, M. Giamberini, *Liquid Crystalline Polymers*. Pergamon Press, Oxford (1994), p 69.

[41] C. Ortiz, L. Belenky, C.K. Ober and E.J. Kramer, *J. Mater. Sci.* 35, (2000), p 2079.

[42] C. Ortiz, R. Kim, E. Rodighiero, C.K. Ober and E.J. Kramer, *Macromolecules*, 31, (1998), p 4074.

[43] H.J. Sue, J.D. Earls and R.E. Hefner, Jr., *J. Mater. Sci.* 32, (1997), p 4031.

[44] Holubka, J.W., Dickie, R.A., Cassatta, J.C., *J. Adhesion Sci. Technol.*, 6(2), 243, (1992)

Shear Flow Effects on Polymer Melts Crystallization: Kinetics Features

*Lucia Bove and Maria Rossella Nobile**
**e-mail: nobile@dica.unisa.it*
Department of Chemical and Food Engineering, University of Salerno, I-84084, Fisciano (Salerno)

Summary: In this work the kinetics of the isothermal crystallization from the melt of isotactic polyolefins in quiescent conditions as well as *after the application of a step-shear flow* is investigated by means of rheological measurements. It is shown that the kinetics of the crystallization, as measured by the increase of the storage modulus, is not affected by the strain amplitude and the frequency of the oscillation, once they are properly chosen.
A strong enhancement of the crystallization kinetics has been obtained when *the step-shear flow* was applied at the crystallization temperature T_C=92°C for two different molecular weight poly(1-butene) samples (i-PB400 and i-PB200) and at T_C=137°C for a polypropylene (i-PP). In particular, the overall-crystallization rate constants of the i-PB400 increased with increasing the applied shear rate at a constant total strain of 60. At higher shear flow temperatures slower kinetics occurred in all the cases until the effect of the applied shear flow was lost. Moreover, the effect of the molecular weight on the flow induced crystallization phenomenon is investigated on the two i-PB samples and the results have clearly shown that the higher molecular weight i-PB200 polymer is much more sensitive than the i-PB400 to the flow history.

Keywords: isothermal crystallization, quiescent, shear-induced, rheology, storage modulus, polypropylene, poly(1-butene).

Introduction

In most processing operations thermoplastic polymers are molten, shaped and solidified under the action of complex flow fields leading, for semi-crystalline polymers, to the phenomenon known as flow-induced crystallization (FIC) that significantly affects the overall-crystallization kinetics [1-6]. Indeed, in polymer melts also a weak shear flow can accelerate the overall kinetics that is enhanced whenever crystallization appears during shear or after shear. Viscoelastic measurements have recently received considerable attention as a capable method to study the crystallization of semi-crystalline polymers either in quiescent conditions [7,8] as well as after the application of a shear flow [3, 9-11]. In particular the storage modulus, G', undergoes an increase of orders of magnitude during the crystallization process because of its sensitivity to the structural changes and therefore it can monitor the crystallization from the melt to the solid

 CCC 1022-1360/00/$ 17.50+.50/0

state.

The transformed crystallized fraction, $\alpha(t)$, can be obtained from the time dependent G'(t) values according to the equation [8]:

$$\alpha(t) = \frac{G'(t) - G'_0}{G'_\infty - G'_0} \qquad (1)$$

where G'_0 and G'_∞ correspond to the G' values of the initial and final plateau, respectively. The α (t) values, ranging between 0 and 1, allow to define the half-crystallization time, $t_{0.5}$, that represents the time at which the crystallized fraction of the sample is 50%. An overall-crystallization rate constant, k, corresponding to the inverse of the half-crystallization time, can be defined from the well known Avrami equation.

In this work the storage modulus has been used as a probe to follow the crystallization process and it will be shown that the crystallization kinetics is not affected by the strain amplitude and the frequency of the oscillation, once they are properly chosen. The *quiescent* isothermal crystallization of two commercial isotactic poly(1-butene) samples of different molecular weight and of a polypropylene sample has been investigated. The results, in terms of the overall-crystallization rate constant, represent the reference condition for the study of the effect of the application of a *step-shear flow*. In these experiments the influence of the shearing temperature, T_S, the applied shear rate and the molecular weight will be analyzed and related to the relaxation behavior of the polymers.

Experimental Part

Materials and methods

In this study two isotactic poly(1-butene) samples provided by *Basel* (Louvain la Neuve), the i-PB400 and the i-PB200, and a standard isotactic polypropylene provided by *Basel* (Ferrara), the i-PPT30G, have been investigated. They are semicrystalline polymers with a melting temperature, T_m, equal to 126°C for the i-PB samples and to 165°C for the i-PPT30G. The molecular weight averages as obtained from GPC measurements by *Basel* are reported in Table1 together with the tacticity as measured by NMR C^{13}.

Sample	M_n	M_w	M_z	M_z/M_w	tacticity (%mmmm)
i-PB400	81700	244000	505000	2.1	79.5
i-PB200	107000	525000	1370000	2.6	82.7
i-PPT30G	55600	376095	1832915	4.9	87.6

Table 1. Molecular characteristics of the poly(1-butene) and polypropylene samples.

It is noteworthy that the two poly(1-butenes) samples used in this work are good candidates to investigate the effect of the kinetics enhancement in the flow induced crystallization due to the increase of the molecular weight. The i-PB200 is, indeed, characterized by the weight average molecular weight, M_w, and the z-average molecular weight, M_z, much higher than the corresponding ones of the i-PB400 sample. Moreover, the ratio M_z/M_w is quite similar for both sample showing a similar molecular weight distribution, MWD, in the high molecular weight tail; i.e. in the MWD region of interest for a flow induced crytallization point of view.
A Rheometrics RDSII rotational rheometer (*Rheometrics Inc.*), equipped with a 2÷2000 g·cm transducer, has been used for both the rheological and the crystallization experiments under nitrogen atmosphere in order to prevent the oxidation of the material.
The rheological characterization of the i-PB and i-PP melt was performed in a parallel plates geometry (∅= 25mm, gap=1mm). In the dynamic oscillatory tests the angular frequency, ω, varied from 0.1 to 500 rad·s^{-1}. The dynamic measurements have been performed at temperatures in the range 120÷190°C in the case of the poly(1-butene) samples and in the range 160÷260°C in the case of the i-PPT30G. Indeed, the rheological measurements started after the samples were annealed in the rheometer between the tools at T=160°C for 10 minutes for the i-PB and at T=240°C for 30 minutes for the i-PP samples to obtain isotropic melts. On the basis of the time-temperature superposition principle the isotherms were shifted to obtain master curves at the reference temperature of T_0=150°C and T_0=230°C for the i-PBs and i-PP polymers respectively.

Rheology of crystallization

The isothermal crystallization experiments were performed adopting a cone-and-plate configuration (∅= 25mm, cone angle= 0.1 rad). The use of the cone-and-plate guaranteed a steady simple shear flow during the application of the *step-shear flow*.

First, experiments in *quiescent* conditions have been performed and the following thermomechanical history has been adopted:

a) the i-PP samples were placed in the rheometer and kept for 30 min at 240°C; the i-PBs samples were kept between the rheometer tools for 10 minutes at 160°C. In both cases the annealing process was performed at a temperature well above the melting temperature (previously reported) to erase crystalline fragments that could affect the subsequent crystallization process;

b) the samples were, then, cooled to the crystallization temperature (T_C) with a rate of about 25°C/min. The crystallization temperatures were T_C= 135°C and 137°C for the i-PP sample, and T_C=92°C for the i-PB samples. The temperature profiles were monitored using a thermocouple in contact with the lower plate in order to minimize temperature differences between the tool and the sample. At the start of the cooling process the evolution of the storage modulus, G', was monitored versus time so that both the cooling step and the isothermal crystallization were analyzed by an oscillatory test at the constant strain amplitude of 1% and at the frequency of 1 rad/s. Indeed, the choice of the values for the strain amplitude and the frequency parameters is crucial in order to reduce as much as possible the disturbance during the crystallization process. These values also ensure that the whole crystallization process occurs in the range of sensitivity of the equipment. To verify whether these parameters could affect the crystallization kinetics, tests with frequency values ranging from 1 to 20rad/s and strain amplitude values ranging from 0.5 to 5% have been performed on the i-PB400 sample.

It is noteworthy that the storage modulus represents our probe to follow the crystallization process.

To study the *flow induced crystallization* (FIC) phenomenon, after the annealing in the melt, a *step-shear flow* at a low shear rate, $\dot{\gamma}$, was imposed during a shearing time, t_s at various temperatures, T_s, in the range between the annealing and the crystallization temperature. Then, after the flow has been stopped, the melt has been cooled at the crystallization temperature and the crystallization was recorded in **isothermal conditions** (at T_C) following G' as a function of time (as previously described). The effect of the flow on the crystallization kinetics was analyzed in terms of the overall-crystallization rate constant, k, corresponding to the reciprocal of the half time of crystallization, $t_{0.5}$.

Results and Discussion

The master curves at the reference temperature of T_0=150°C for the i-PB400 and the i-PB200 samples and at T_0=230°C for the i-PP polymer have been obtained with the time-temperature superposition principle. In Figure 1 the master curve relative to the i-PB400 is reported in terms of the elastic and loss modulus (G' and G" respectively).

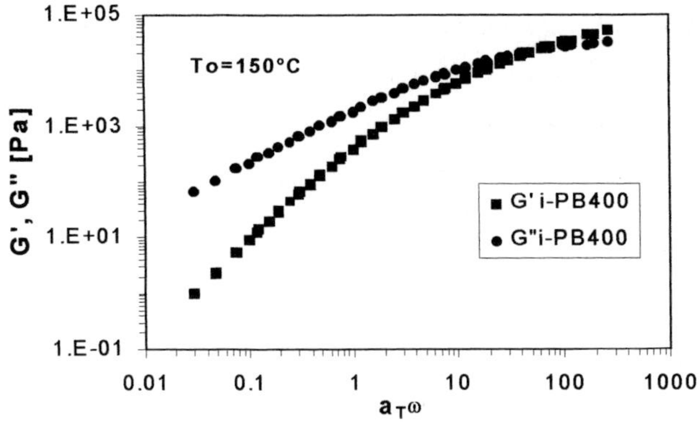

Figure 1. Master Curve of the i-PB400 at T_0=150°C.

The corresponding shift factors, a_T, are plotted vs 10^3 (1/T) (K^{-1}) and fitted with the Arrhenius expression, as shown in Figures 2a, b ,and c for the three polymers investigated.

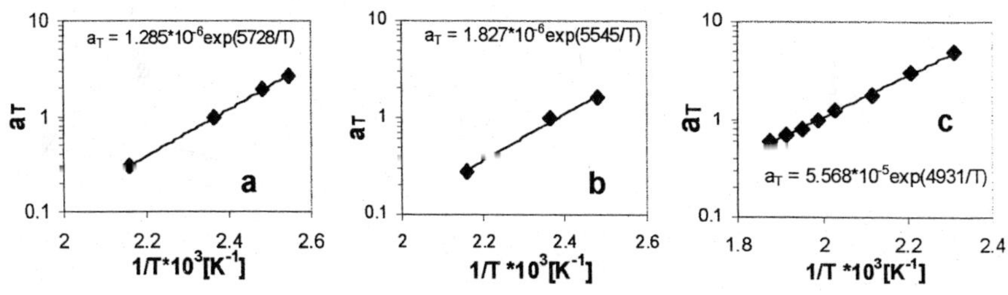

Figure 2. Shift factors a_T, vs 10^3 (1/T) (K^{-1}) for: **(a)** i-PB400; **(b)** i-PB200; **(c)** i-PPT30G.

The relationships between a_T and (1/T) (K^{-1}), reported in the Figure 2, are then used to evaluate the relaxation time spectra, H(τ), of these polymers at different temperatures (see Figures 3a, b and c) once H(τ) is calculated at the reference temperature from the master curves. To estimate the reference H(τ), first the dynamic moduli have been related to the relaxation moduli (as obtained from the GEX-double reptation modulus) using a Finite Element Approximation and then the Aury law was used. The details of the calculus of H(τ) are reported in a previous paper by Nobile and Cocchini [(12)]. It is noteworthy that this procedure gives exactly the same H(τ) that has been more recently evaluated by Nobile and Cocchini [(13)] using the analytical relationship connecting the molecular weight distribution and the relaxation spectrum proposed by Thimm et al. [(14)]

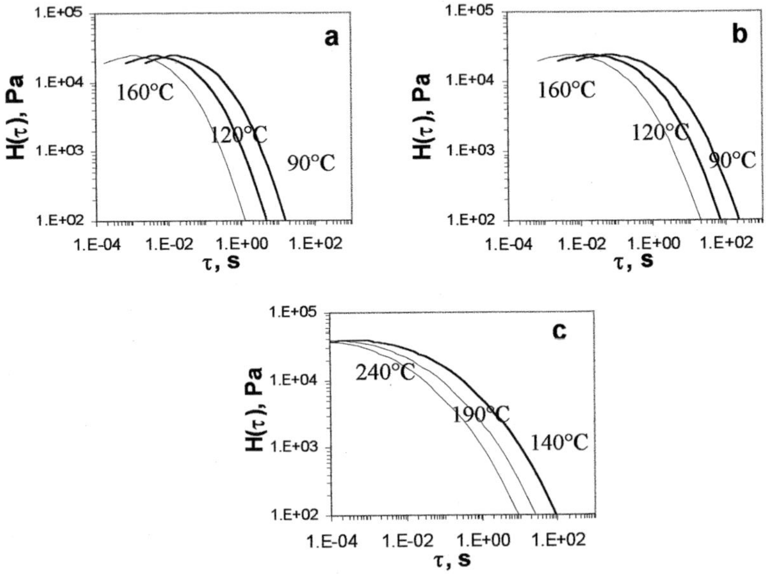

Figure 3. Relaxation time spectra, H(τ), for: **(a)** i-PB400 and **(b)** i-PB200 at T=160, 120, 90°C; **(c)** i-PPT30G at T=240, 190,140°C.

The comparison between Figures 3a and 3b shows that the i-PB200 is characterized by much longer relaxation times than the i-PB400, as expected due to its the higher molecular weight (see Table 1). This event again confirms that the two i-PB polymers chosen in this study represent good candidates for the study of the effect of molecular weight on the flow induced crystallization phenomenon since a major factor governing FIC is the relaxation behavior of the polymer chains.

Rheology of the isothermal quiescent crystallization

As previously mentioned in the Experimental section, we have first verified that the use of the storage modulus represents a reliable probe to follow the crystallization kinetics process. The analysis of the influence on the crystallization process of the applied strain amplitude has been performed on the i-PB400 sample by varying the strain amplitude values from 0.5 to 5%. The sample has been crystallized during the oscillatory measurements at Tc=92°C with a constant frequency ω=1 rad/s .

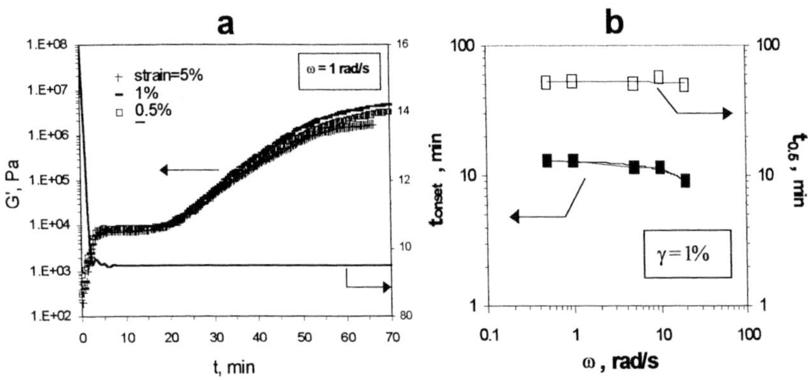

Figure 4. Quiescent crystallization of the i-PB400 at T_C=92°C monitored with oscillatory experiments. **(a)** Evolution of G' vs time applying a constant frequency (ω=1 rad/s) and different strain amplitude values (γ = 0.5÷5 %); **(b)** onset times, t_{onset} , and half crystallization times, $t_{0.5}$, evaluated from G' data, applying a constant strain amplitude (γ=1%) and different frequencies (ω=0.5÷20rad/s)

In Figure 4a the evolution of G' vs. time during both the cooling from the melt and the isothermal crystallization is reported together with the temperature profiles. After a first increase of the modulus during the cooling step, a plateau (G'_o) at T_C is observed, followed, after an instrumental onset time, t_{onset}, by a second strong growth of the modulus representative of the crystallization phenomenon and by a final plateau (G'_∞) at the end of the process. The results show that no evident effects on the crystallization kinetics of the applied strain amplitude, γ , in the range 0.5%÷5% can be detected. Therefore all the dynamic measurements reported in the following have been performed with a strain amplitude of 1%.

In Figure 4b the effect of the applied frequency is shown in terms of the onset and half-crystallization times. Again the results show that the applied frequency has no significant effect on the quiescent crystallization kinetics monitored with oscillatory tests, in the frequency range investigated. Indeed, only a slight influence on the onset time value is observed at the highest frequencies. The frequency of 1 rad/s has been, then, chosen to perform all the crystallization experiments.

Rheology of the isothermal crystallization after a step-shear flow

The influence of the *step-shear flow* applied at different temperatures, T_s, on the overall crystallization kinetics has been investigated for the i-PB200 and i-PB400 polymers at Tc=92°C. The results are reported in Figure 5 in terms of the ratio between the overall-crystallization rate constant as measured in the G' crystallization experiments *after the step-shear flow*, k_S, and the corresponding value of the quiescent experiment, k_0. The quiescent experiment is, indeed, our reference condition and the effect of the *step-shear flow* on the crystallization kinetics will be always reported in comparison to the quiescent case.

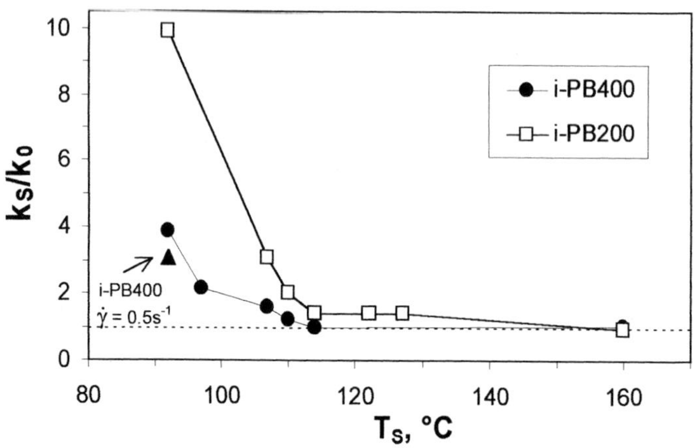

Figure 5. The effect of the shearing temperature on the overall crystallization rate at Tc=92°C for the i-PB200 sample (shearing conditions: $\dot{\gamma}$=0.5s^{-1}, γ=60) [open squares] and i-PB400 sample (shearing conditions: $\dot{\gamma}$=1s^{-1}, γ= 60 [full circles] and $\dot{\gamma}$=0.5s^{-1}, γ= 60 [full triangle]).

When a *step-shear flow* with the shear rate of 1 s^{-1} and γ= 60 is applied to the i-PB400 at the crystallization temperature of 92°C the overall isothermal crystallization kinetics is strongly accelerated, i.e. the k_S/k_O is 3.9. Increasing the temperature at which the shear flow is applied, T_s, from 92°C to 114°C, slower kinetics occur. For T_s higher than 114°C the effect of the applied shear flow is lost (i.e. the ratio k_S/k_O becomes 1).

These results clearly show that a major factor governing the flow induced crystallization phenomenon is the relaxation behavior of the polymer chains. Indeed, the relaxation spectrum of the i-PB400 polymer, reported in Figure 3a, shows a significant increase of the longer relaxation times with decreasing the temperature from 160°C to 90°C. In the case of the higher molecular weight sample, the i-PB200, the effect of *the step-shear flow* has been investigated at the shear rate of 0.5 s^{-1} that was the highest possible shear rate at the lower T_s used (due to the transducer limit). Also in this case a total strain, γ, equal to 60 was adopted for all the experiments. Again, the effect of the applied shear flow vanishes increasing the temperature T_s.

A strong influence of the molecular weight on the FIC, as expected, is evident from our results. Indeed, in order to compare exactly the same shear flow condition, the result for the shear flow applied to the i-PB400 at Tc=92°C with a $\dot{\gamma}$= 0.5s^{-1} and γ=60 is also reported in Figure 5. The k_S/k_O value for the i-PB400 is 3.1 and for the i-PB200 is 9.9 showing that the higher M_w sample is much more sensitive to the shear flow history, as also clearly reported in the works by Somani *et a.l* [(4)] and by Jay *et al.* [(6)]. Again, this results can be justified in terms of the relaxation behavior of the polymers. The relaxation spectra of the i-PB200 sample (Figure 3b) are, in fact, characterized by relaxation times much longer than the corresponding ones of the i-PB400 (Figure 3a).

To investigate the effect of the applied shear rate, step-shear flow experiments have been performed on the i-PB400 sample at the crystallization temperature of 92°C with shear rates ranging between 0.25 and 5 s^{-1} and a constant total strain γ= 60. In Figure 6a the results for two tests at $\dot{\gamma}$=0.25s^{-1} and $\dot{\gamma}$=5s^{-1} are reported in terms of G' vs. time together with the results for the quiescent crystallization. A very good reproducibility is obtained in all the rheological crystallization experiments. The results of the analysis of the applied shear rate in the whole range investigated is shown in Figure 6b in terms of the overall-crystallization rate ratio, k_S/k_O.

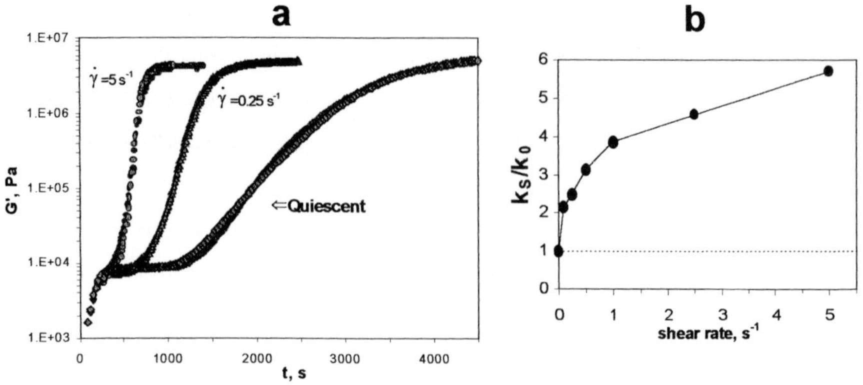

Figure 6. Shear rate effects for the i-PB400 crystallized at T_C= 92°C after a *step-shear flow* at T_S= 92°C and γ = 60: **(a)** G' vs time after the *step-shear flow* for two tests with $\dot{\gamma}$=0.25, two tests with $\dot{\gamma}$=5 s^{-1} and two tests for the quiescent crystallization kinetics; **(b)** k_S/k_O vs the shear rate in the range $\dot{\gamma}$=0.1÷5 s^{-1} investigated.

At a fixed shear deformation, the higher is the applied shear rate, the faster the crystallization occurs after the flow has been stopped.

In Figure 7 the results of the study of the *flow induced crystallization* of the i-PPT30G with different shear rates, $\dot{\gamma}$= 1, 2.5 and 5 s^{-1} at the same total deformation of γ=1000 are reported.

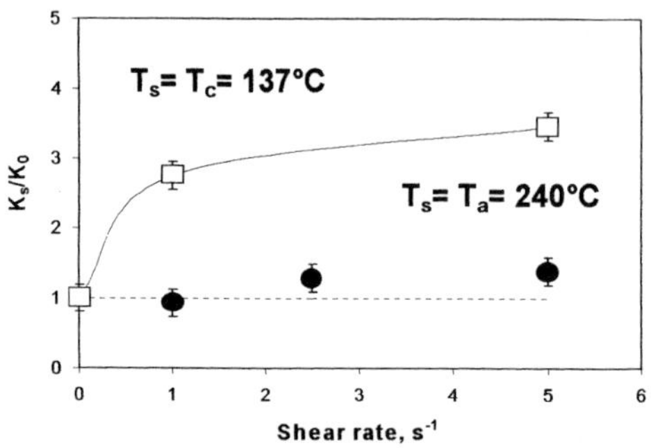

Figure 7. Effect of the shear rate on the crystallization rate constant of the i-PP T30G. Filled symbols: shearing temperature, T_s= 240°C; shearing deformation, γ=1000; crystallization temperature, T_c = 135°C. Open symbols: T_s=T_c=137°C, γ =200.

It is possible to observe that a *step-shear flow* imposed in the melt at the shearing temperature

T_S= 240°C (filled circles in Figure 7) has a very low effect on the subsequent crystallization kinetics: the ratio between the sheared and the quiescent overall-crystallization rate constants, k_S/k_O, does not exceed 1.4 , even if the applied strain, γ =1000, is quite high. In order to produce stronger effects the shear flow must be imposed at much lower temperatures, *e.g.* at the crystallization temperature (T_C=137°C). In these conditions the shear flow is highly effective and the overall-crystallization rate constant is seen to increase, at constant shear deformation, with increasing the shear rate (open squares in Figure 7).

Conclusions

In this work the *quiescent* isothermal crystallization of two different molecular weight isotactic poly(1-butene) polymers and of an isotactic polypropylene sample, as well as the crystallization *after the application of a step-shear flow*, has been studied by means of rheological measurements. In particular, it was shown that the kinetics of the crystallization as measured by the increase of the storage modulus is not affected by the strain amplitude and the frequency of the oscillation, once they are properly chosen.

A strong enhancement of the crystallization kinetics has been obtained for all the samples when *the step-shear flow* was applied at the crystallization temperature. In particular, the overall-crystallization rate constants increased with increasing the shear rate applied at a constant total strain of 60. At higher shear flow temperatures slower kinetics occurred until the effect of the applied shear flow was lost. Moreover, the results have shown that the higher molecular weight i-PB200 polymer is much more sensitive to the flow history than the i-PB400. These results clearly show that a major factor governing the flow induced crystallization phenomenon is the relaxation behavior of the polymer chains.

Acknowledgments

The financial support of the Italian Ministry of University and Scientific Research, PRIN 1999-2001,"Flow-induced Crystallization of Polymers. Impact on Processing and Manufact Properties", is gratefully acknowledged.

References

1. G. Kumaraswamy, A.M. Issaian and J.A. Kornfield, *Macromolecules*, 32, 7537-7547 (1999)
2. A.K. Doufas, I.S. Dairanieh and A.J. McHugh, *J. Rheol.*, 43, 85-109 (1999)
3. C. Tribout, B. Monasse and J.M. Haudin, *Colloid Polym. Sci.*, 274, 197 (1996)
4. R.H. Somani, B.S. Hsiao, A. Nogales, S. Srinivas, A.H. Tsou, I. Sics, F.J. Balta.Calleja and T.A. Ezquerra, *Macromolecules*, 33, 9385-9394 (2000)
5. N. Pogodina and H.H. Winter, *Macromolecules*, 31, 8164-8172 (1998)
6. F. Jay, J.M. Haudin and B. Monasse, *J. Mater. Sci.*, 34, 2089-2102 (1999)
7. K. Boutahar, C. Carrot and J. Guillet, *Macromolecules*, 31, 1921-1929 (1998)
8. Y.P. Khanna, *Macromolecules*, 26, 3639-3643 (1993)
9. S. Vleeshouwers and H.E.H. Meijer, *Rheol. Acta*, 35, 391-399 (1996)
10. L. Bove, E. Somma and M.R. Nobile, *Mat. Eng*, 3 (2001), in press
11. L. Bove, M.R. Nobile, F. Azzurri and G.C. Alfonso, *Proceedings of the PPS 17, Canada,* May 21-24 (2001)
12. M.R. Nobile and F. Cocchini, *Rheol. Acta*, 40, 111-119 (2001)
13. M.R. Nobile and F. Cocchini, *Proceedings of the 2nd Workshop on Inverse Problem in Rheology, Germany,* May 9-11 (2001)
14. W. Thimm, C. Friedrich and T. Roth, *J. Rheol.*, 44, 1353-1361 (2000)

Macromol. Symp. ***180,*** *181–189 (2002)*

Study of the Influence of Atmospheric Pollutants on the Natural Ageing of Rigid Polyvinyl Chloride

Naïma Belhaneche- Bensemra, Naïma Ouazene

Laboratoire de recherche en sciences et techniques de l'Environnement, Ecole Nationale Polytechnique, BP 182, El-Harrach, Alger, ALGERIE
(E- mail: nbelhaneche@yahoo.fr)

Summary: The aim of this work is to study the natural ageing of rigid polyvinyl chloride (PVC) used in building and construction areas. For that purpose the PVC bars have been exposed in Algiers for 24 months in three sites where the concentrations of atmospheric pollutants (NO_x, O_3, hydrocarbons) are known. Samples were taken every three months. The evolution of mechanical properties was followed by tensile and hardness tests whereas the modification structure of polymer was followed by Fourier transform infrared spectroscopy (FTIR) and UV-visible spectroscopy.
The results have shown a little drop in strain at break and stress at break, whereas hardness remains practically constant.
IR and UV- visible analysis have shown apparition of chemical structures due to the polymer degradation (carbonyl and hydroperoxide groups and polyenes) since the first months of exposure in the three sites.

Introduction

Most commercial polymers undergo chemical and physical modifications when they are exposed for a long time to the combined action of sunlight, atmospheric oxygen, moisture and pollutants [1].

Air pollution is usually defined as the presence in the outdoor atmosphere of substances put there directly or indirectly by an act of man, in amounts which are detrimental to the health and safety or interfere with the full use of materials or property, e.g. made from polymeric materials [2]. Man is not the only agent able to pollute the atmosphere. There are many natural processes that do so such as pollination of plants, volcanic eruptions, dust storms and forest fires. There are also secondary pollutants in the atmosphere, formed in the air from primary pollutants, due to acts by man, e.g. smoke, industrial pollutants and photochemical smog formed in the air from substances emitted from automobile, exhaust and other sources. The major air pollutants are particulate matter, sulfur oxides, nitrogen oxides, hydrocarbons and photochemical oxidants such as ozone, atomic oxygen, singlet oxygen, and peroxyacetylnitrate [3].

The concentration of air pollutants is very important for the long term properties of polymeric materials. Polyvinyl chloride (PVC) is a polymer of considerable commercial importance.

 CCC 1022-1360/00/$ 17.50+.50/0

A wide variety of applications in outdoor constructions such as doors, down spouts, shutters, sash window, exterior cladding and advertising signs, where impact resistance, transparency, color, corrosion resistance and water repellency are of extreme considerations have been developed. However PVC is very susceptible towards discoloration during as well as after the processing of polymer. When PVC is exposed to long term outdoor environment, it undergoes very slow long term photochemical degradation. It is subjected to various atmospheric effects including exposure to rain, fungus, mechanical stress, pollutants and exposure to ultraviolet radiation. The ultimate result is then reflected in a marked loss of additives, discoloration, embrittlement and finally loss in mechanical properties [4].

The aim of this paper is to study the natural weathering of rigid PVC used in building and construction areas. For that purpose, the PVC bars have been exposed in Algiers for 24 months in three sites were the concentrations of atmospheric pollutants (NO_x, O_3 and hydrocarbons) are known. Samples were taken every three months. The evolution of mechanical properties was followed by tensile and hardness tests whereas the modification structure of polymer was followed by Fourier transform infrared spectroscopy (FTIR) and UV-visible spectroscopy.

Experimental

- **Materials**

This study of weatherability was carried out on industrial samples manufactured by ENPC-PROFIPLAST (ALGERIA). Commercial grades of additives listed in Table 1 were used in the formulation. The percentages of components were not given by the producer.

This rigid PVC is used to produce various articles such as doors, windows and shutters.

Table 1. Compounds used.

Compound	Chemical nature
Heat and light stabilizer	Lead bibasic phosphite
External lubricant	Lead stearate
Impact modifier	Acrylic resin
Pigment	TiO_2
Filler	$CaCO_3$

- **Exterior Weathering in Algiers**

The samples (100×6×4 cm) were exposed to exterior weathering according to the standard ISO 4607 (1985).

They were deposited in spring 1998 for two years in three sites where the concentrations of atmospheric pollutants are known:

Site I:

It is representative of an urban area.The amount of aromatic hydrocarbons is about 140 μg/m^3.

Site II:

It is situated in the center of the town. This stuffy site is characterized by a heavy traffic (1400-1700 vehicles/day) which permits the accumulation of pollutants. The levels of NO_x are high (NO and NO_2 average contents/hour are respectively 10-100 PPb and 10-60 PPb).

Site III:

It is situated in the suburbs and is characterized by a weak traffic. The level of O_3 is high (O_3 average content/hour = 20-90 PPb).

The weathered samples were taken every three months and characterized. On the other hand, samples were keeped in an interior ambience to serve as reference. They were characterized in the same manner than the weathered ones.

- **Mechanical Characterization**

Measurements of tensile properties were undertaken using an ADAMEL LHOMARGY DY 25 testing machine according to the AFNOR NF T51-034. A shore D type durometer was used for determining the hardness of the samples according to the AFNOR NF T51-109.

- **Spectroscopic Characterization**

Before spectroscopic characterization a sample was cut in a cross- section of the whole bar and was purified by dissolution in tetrahydrofuran (THF), centrifugation, filtration, precipitation with methanol and then filtration and drying in vacuum at 40°C during 24 hours. The infrared spectra were recorded with a Philips type PU 9800 FTIR spectrophotometer at a resolution of 2 cm^{-1} using KBr pellets (1wt %).

The UV visible spectra were recorded with a SECOMAM ANTHELIE LIGHT UV visible spectrophometer. Solutions of the samples were prepared in freshly distilled THF (1 g/l) and analyzed immediately.

Results and Discussion

Figures 1 and 2 illustrate the variation of stress and strain at break as a function of the time of exposure in months, respectively. They show clearly that the properties at break in the three sites decreased with exposure time while they remained practically constant for the reference. Although the variations were relatively weak, the effect of exposure site was evidenced. It seems that the O_3 exerted the most nocuous effect, followed by hydrocarbons and then NO_x.

The decrease of the mechanical properties at break can be related to the chains scissions due to the photooxidation of PVC [5].

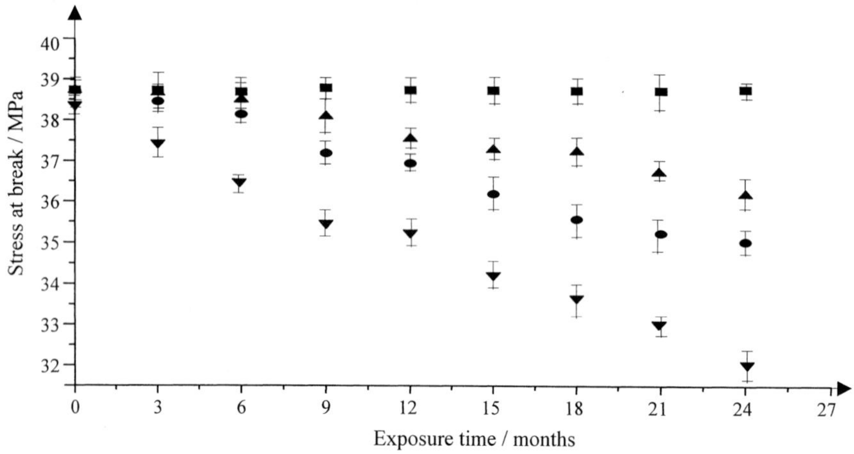

Figure 1. Effect of exposure site on the evolution of stress at break :
■ Reference, ● Site I, ▲ Site II, ▼ Site III

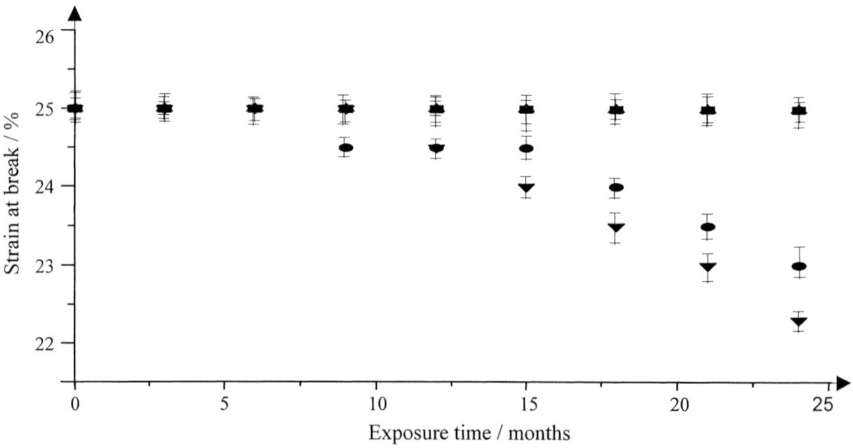

Figure 2. Effect of exposure site on the evolution of strain at break :
■ Reference, ● Site I , ▲ Site II, ▼ Site III

The Shore D hardness values obtained for all the weathered samples in the three sites are reported in Table 2. It can be noted that hardness remained practically constant for all the samples and in the three sites. The considered time exposure and the low level of degradation have not affected the hardness of the samples.

Table 2. Shore D hardness.

Time (months)	00	03	06	09	12	15	18	21	24
Reference	75	75	75	75	75	75	75	75	75
Site I	75	75	75	75	74	75	75	75	75
Site II	75	75	74	74	75	75	75	75	75
Site III	75	75	75	75	74	75	74	75	75

It is known that several complex chemical changes take place in PVC during weathering, including dehydrochlorination and oxidation which in turn lead to changes in color (both coloration and bleaching), crosslinking and chain scission, and general deterioration of physical properties [1,4]. FTIR and UV-visible spectroscopy were used, after purification of the samples, to evaluate the effects of the degradation in the three sites. The FTIR spectra (Figures 3, 4 and 5), showed significant changes with ageing. A carbonyl peak centered around 1733 cm^{-1} and a very broad hydroxyl peak at 3200-3600 cm^{-1} developed after exposure in the three sites since the first months.

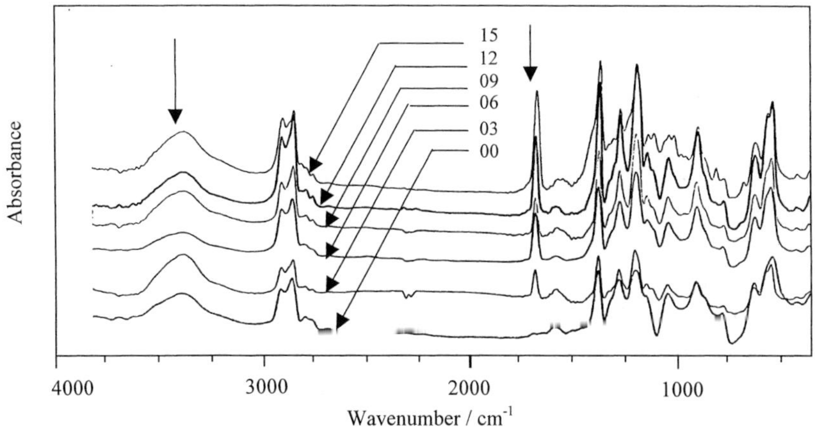

Figure 3. Evolution of FTIR spectra with exposure time (months) in site I.

These different types of oxygenated structures appear in the PVC chain on photolysis in the presence of oxygen [1,2]. The very broad hydroxyl band can be attributed to hydroperoxide and alcohol functions [1,6]. The carbonyl band was suggested to be mainly due to α and β-chloroketones [1,6].

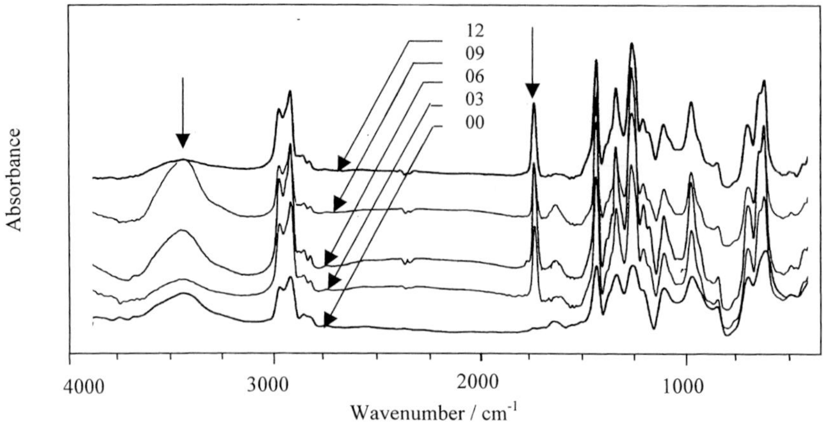

Figure 4. Evolution of FTIR spectra with exposure time (months) in site II.

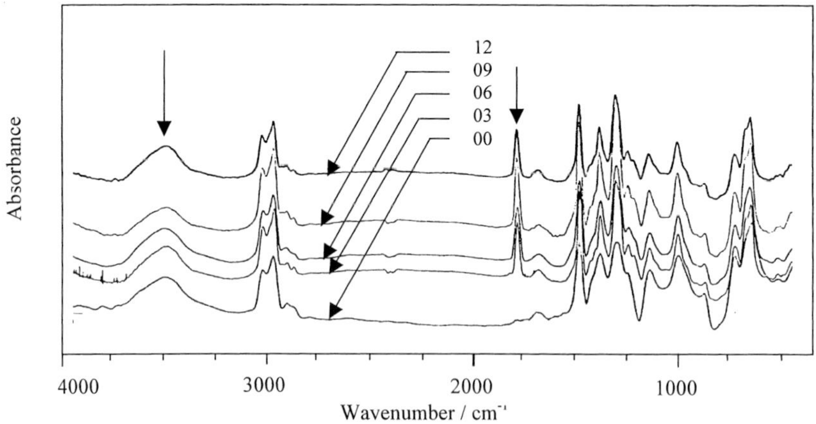

Figure 5. Evolution of FTIR spectra with exposure time (months) in site III.

In Tables 3 and 4 are reported the relative variations of carbonyl and hydroxyl bands, respectively, with the exposure time. The reference band at 1434 cm^{-1} is due to the $-CH_2-$ groups of PVC. It can be seen that the more rapid increase of the ratios A_{1733} / A_{1434} and A_{3430} /A_{1434} occured during the first months of exposure. After that, the evolution of these ratios is relatively weak. These results indicate that the oxidation process develop primarily in the top layer of the weathered material. This leads to the formation of oxygenated structures which absorb the UV light and then protect the deep layers.

Table 3. Evolution of the ratio A_{1733}/A_{1434} with exposure time.

Time (months)	00	03	06	09	12	15
Site I	00	0.61	0.67	0.70	0.69	0.72
Site II	00	0.59	0.60	0.60	0.64	0.66
Site III	00	0.59	0.63	0.65	0.66	0.74

Table 4. Evolution of the ratio A_{3430}/A_{1434} with exposure time.

Time (months)	00	03	06	09	12	15
Site I	0.08	0.32	0.32	0.35	0.41	0.45
Site II	0.08	0.10	0.20	0.61	0.10	0.15
Site III	0.08	0.94	0.23	0.21	0.24	0.34

On the other hand, it is known that the main characteristic feature of the photolysis of PVC is the dehydrochlorination reaction, which develops extensively, as in the thermal degradation, and which leads to the formation of conjugated polyene sequences with concomitant evolution of HCl [1,7,8]. These polyene structures can be easily characterized by their UV-visible absorption, which ranges from 230 nm for dienes to ≈ 400 nm for n=8 [9]. Figures 6, 7 and 8 shows the UV-visible spectra obtained after purification of the aged samples in the three sites. Similar spectra were given in the literature on the photooxidation of PVC [10-12].

Contrary to the non aged PVC which has no absorption band, all the aged samples presented different absorption bands below 350 nm. These bands are representative of the formation of conjugated polyenes with $n \leq 6$ [9].

Furthermore, the corresponding absorbances increased with the exposure time in the three sites indicating the increase of the concentration of these various polyenes. As degradation proceeds, these chromophores concentrate closer to the surface of the exposed material, leading to the appearance of a highly degraded superficial layer which acts as a UV screen and partly prevents the bulk of the PVC sheet from further deterioration. In the presence of oxygen, oxidation develops simultaneously with the zip-dehydrochlorination with formation of carbonyl and hydroperoxide groups on the polymer chains which in addition suffer cleavages and crosslinking leading to the deterioration of mechanical properties.

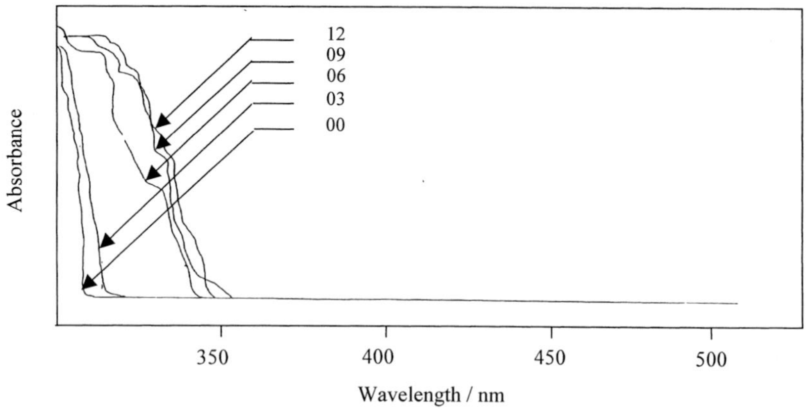

Figure 6. Evolution of UV-visible spectra with exposure time (months) in site I.

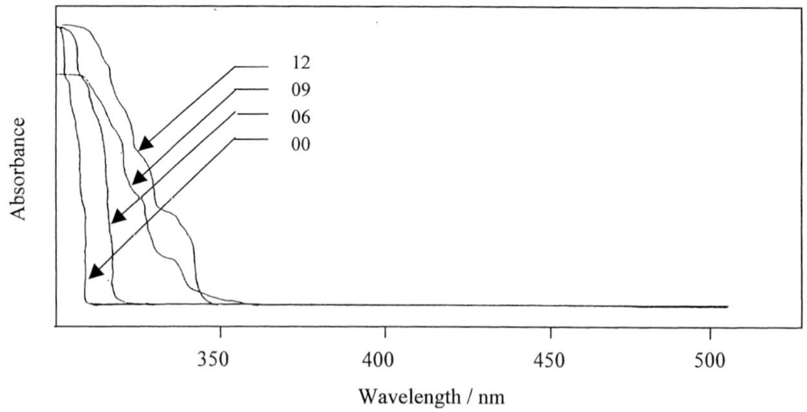

Figure 7. Evolution of UV-visible spectra with exposure time (months) in site II.

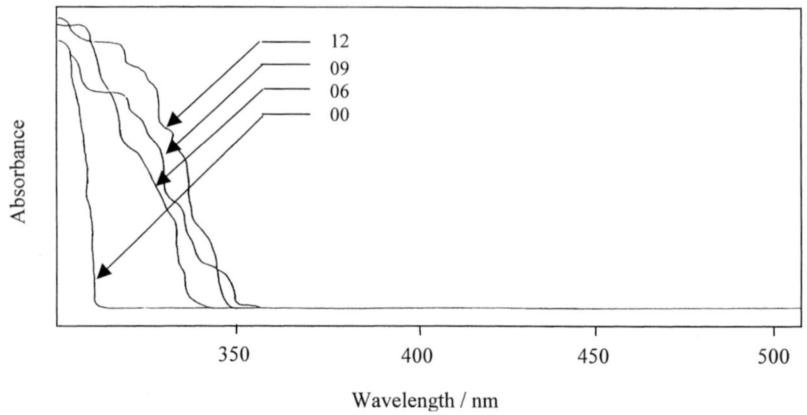

Figure 8. Evolution of UV-visible spectra with exposure time (months) in site III.

Conclusion

The investigation of the influence of atmospheric pollutants(O_3, NO_x, hydrocarbons) on the weatherability of rigid PVC was carried out. The samples were exposed for two years in three sites. Every one of them was characterized by a relatively high level of a specific atmospheric pollutant.

The results showed a decrease in the properties at break of the samples since the first months of exposure. It was found that O_3 exerted the most deleterious effect, followed by hydrocarbons and then NO_x. However, shore D hardness remained practically constant for all the samples in the three sites. The considered time exposure and the level of degradation have not affected the hardness of the samples.

FTIR and UV-visible analysis showed the formation of chemical structures due to the polymer photoxidation (carbonyl and hydroperoxide groups and polyenes) since the first months of exposure. During the oxidation, chains scissions and crosslinking reactions occurred leading to the gradual loss of the mechanical properties .

This paper evidenced the influence of air pollutants on the long-term properties of PVC. The need of more research is amply verified, both on the effects and their prevention.

[1] C. Decker, "Degradation and stabilization of PVC ", Elsevier Applied Science Publishers, London and New York **1984**, p81.
[2] J. R Hollum,"Topics and terms in environmental problems", Wiley, New York **1978**, p52.
[3] S. Garner, G. Stahl, The effects of hostile environment on coatings and plastics, *American Chemical Society* **1983**, p291.
[4] B.D Gupta, J. Verdu, *J. Polym. Eng* , Nos 1-2, **1988**, 8, 73.
[5] C. Decker, M. Balandier, *Eur. Polym. J.* **1982**, 18, 1085.
[6] G.Scott, M. Tahan, *Eur.Polym.J.* **1975**, 11, 535.
[7] F. Mori, M. Koyama, Y. Oki, Die Angew. *Makromol. Chem.*, **1979**, 75, 113.
[8] J. L. Gardette, J. Lemaire, *Polym. Degr. Stab.* **1989**, 25, 293.
[9] V. D. Daniels, H. H. Rees, *J. Polym.Sci., Polym. Chem.* Ed. **1974**, 12, 2115.
[10] M. Balandier, C. Decker, *Eur. Polym . J.* **1978,** 14, 995.
[11] C. Decker, M. Balandier, *Eur. Polym.* J. **1982**,18, 1085.
[12] J. L. Gardette, J. Lemaire, *Polym. Degr.Stab.* **1986**, 16, 147.

Study of the Migration of Additives from Plasticized PVC

Naîma Belhaneche-Bensemra, Chafia Zeddam, Salah Ouahmed

Laboratoire de recherche en sciences et techniques de l'Environnement,
Ecole Nationale Polytechnique, BP 182, El-Harrach, Alger, Algérie
(E-mail: nbelhaneche @yahoo.fr)

Summary: The aim of this paper is to study the interactions between packaging and food. For that purpose, additives were first characterized. Kinetic studies of their specific migrations have been carried out by using different analytical methods such as gas chromatography, Fourier transform infrared spectroscopy, atomic absorption spectrometry and differential scanning calorimetric analysis. The influence of various parameters such as temperature, agitation, nature of food simulant and initial concentration of plasticizer was considered.

Introduction

Plastics are being used increasingly for packaging foodstuffs, but not all are suitable for this purpose. It is well known that, apart from high polymers, plastics materials also contain low molecular compounds, particularly additives such as stabilizers, lubricants and plasticizers. The addition of such substances is essential for processing and for achieving the desired chemical and mechanical properties[1].

The possible migration of minor constituents from polymeric packaging materials has become increasingly important. The migration from flexible packaging materials in contact with food would not normally include major components of the package itself but may contain minor constituents. Some of the migrants, can and do effect quality of the contained product as exhibited by sensorily determinable changes (odor and / or taste) or by toxicological symptoms from ingestion. The first case is of economic importance, while the second relates directly to health hazards with carcinogens as the worst case[2, 3].

Polyvinyl chloride (PVC) films have found wide applications in the packaging of large variety of foodstuffs, such as fresh meat, fruits and vegetables, and cheeses[4]. These commercial films, in addition to the polymeric component, contain a number of additives.The one used in the highest - proportion is the plasticizer[1].

Plasticizers are low- melting solids or high- boiling organic liquids added to plastic polymers in order to aid flow and processing, to extend and modify the natural properties of the resin and to develop new, technologically important properties not present in the resin itself[5].

 CCC 1022-1360/00/$ 17.50+.50/0

Among the most commonly used plasticizers are di-2 ethyl hexyl phtalate (DEHP) and di-2-ethyl hexyl adipate (DEHA)[6].

The amount of package components that may be leached by food or food simulating solvants depends on the original concentration of the particular component or migrant in the polymer, its solubility in the solvent and / or the partition coefficient between the polymer and solvent as well as temperature and time[7].

The aim of this paper is to study the migration of additives from plasticized PVC. Kinetic studies of their specific migrations have been carried out by using different analytical methods such as gas chromatography , Fourier transform infrared spectroscopy (FTIR), atomic absorption spectrometry (AAS) and differential scanning calorimetric analysis (DSC). The influence of various parameters such as temperature, agitation, nature of food simulant and initial concentration of plasticizer was considered.

Experimental

- **Materials**

Commercial grades of resin and additives listed in Table 1 were used without purification.

Table 1. compounds used in this study.

Compound	Commercial name	Source
Polyvinyl chloride	SHEINTECH	ITOCHU (FRANCE)
di-2-ethyl hexyl phtalate (plasticizer)	Palatinol AH-L	BASF (GERMANY)
Tin-based thermostabilizer	IRGASTAB 17 MOK	CIBA-GEIGY (SWITZERLAND)
Aid -process	FAIR PLAST PA 205	PROMOPACK(ITALY)
Esters of fatty acids (internal lubricant)	Loxiol GH4	HENKEL (GERMANY)
Metal soaps (external lubricant)	Loxiol G78	HENKEL (GERMANY)

Several formulations containing 3% of the processing aid, 4,5 % of the stabilizer, 3,6 % of the internal lubricant, 0,9 % of the external lubricant and, respectively, 30, 40, 50 and 60 % of plasticizer (DEHP) were realized. PVC and additives were mixed in a two-roll mill at 180°C. The specimens for migration tests were melt compressed at 150°C under a pressure of 12 bars during 15 minutes. Then, circular samples of 2 mm in thickness and 19 mm in diameter were cut.

- **Migration Testing**

Migration tests were conducted using four food simulants (sunflower oil, distilled water, 3% (w/v) aqueous acetic acid, 15% (v/v) aqueous ethanol). These food simulants represent all fatty, liquid and moist foods and beverages, except for beverages with a high alcoholic strength. The test conditions used were 10 days at 25 and 45°C with and without agitation. Ten (10) circular samples were immersed in 100 ml of each food simulant. A circular sample and 10 ml of food simulant were taken off every day to be analyzed. Each sample was wiped and weighed. The rate of variation of the mass (τ) was determined as a function of time following the relation:

$$\tau\ (\%) = \frac{m_t - m_0}{m_0} \times 100$$

where: m_0 = initial mass before immersion , m_t = mass of the sample at the time t.

- **Gas Chromatography (GC)**

A Philips PYE Unicam 304 GC with a flame ionization detector was used. Chromatographic conditions were as follows : column OVI of dimensions 1.8 m×4mm o.d ; column temperature : 280°C ; injection temperature: 300°C; detector temperature: 300°C; nitrogen flow: 20.5 ml /min

- **FTIR Analysis**

A Philips type PU 9800 FTIR spectrophotometer was used. The food simulants were placed between two KBr pellets and analyzed directly. On the other hand, the PVC circular samples were dissolved in tetrahydrofuran. After evaporation of the solvent, a polymeric film was recuperated and analyzed.

- **DSC Analysis**

Glass transition temperatures were measured with a Perkin Elmer DSC- 7 apparatus. Each sample was heated from 5 to 110°C under nitrogen with a heating rate of 10°C/ min.

- **Atomic Absorption Spectrometry**

A Philips PYE UNICAM SP 6 spectrometer was used.

Results and Discussion

- **Preliminary Study of the Rate of Mass Variation**

The rates of mass variation (τ) as a function of time gives informations about the phenomenon which occurred between the samples and the food simulants used. An increase means that the food simulant penetrated the sample while a decrease means that some additives migrated in the food simulant. Hence τ gives informations about the overall migration that occurred. Figures 1 and 2 illustrate the variation of τ as a function of time for the four food simulants used at 45 °C with and without agitation, respectively. It can be noted that the highest rates of mass variations were obtained in sunflower oil indicating that migration of additives occurred. For aqueous simulants the rates of mass variations are practically the same and a very weak increase of τ was observed indicating penetration of the food simulant in the PVC discs. On the other hand, the highest values of τ were obtained during the migration tests with agitation and for the formulations containing the highest amounts of plasticizer. This is expected, since the solubility of DEHP is better in sunflower oil than in aqueous simulants. Furthermore, the migration of DEHP can involve the migration of the other additives which are present in the formulation like stabilizer and lubricant.

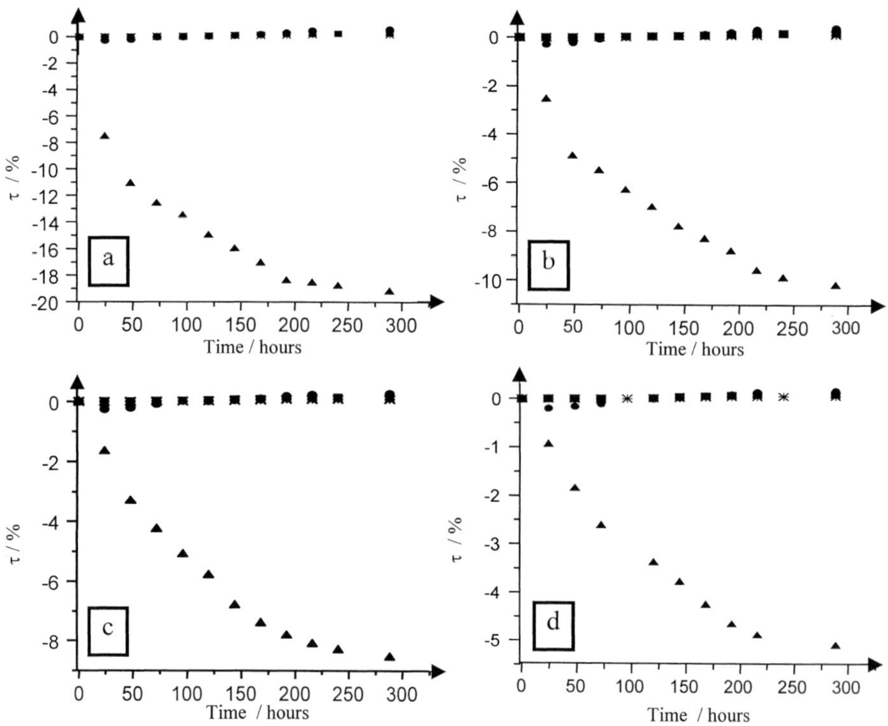

Figure 1. Effect of the nature of the food simulant on the rate of mass variation τ at 45°C with agitation : a) 60 % DEHP ; b) 50 %DEHP ; c) 40 % DEHP ; d) 30 % DEHP :
▲ Sunflower, ● 15 % v/ v aqueous ethanol, ■ 3 % w/v aqueous acetic acid ,
✳ Distilled water

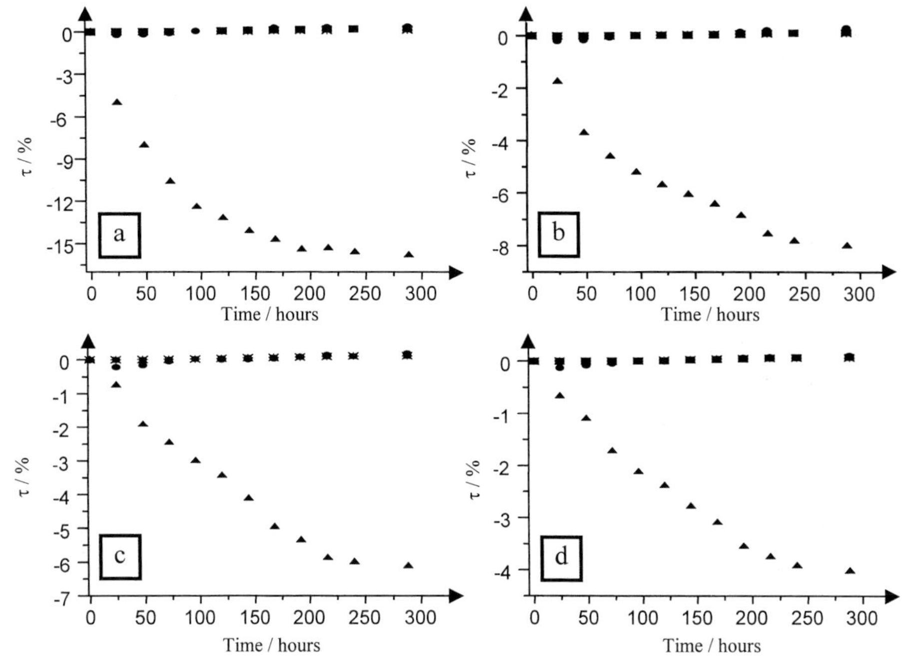

Figure 2. Effect of the nature of the food simulant on the rate of mass variation τ at 45°C without agitation : a) 60 % DEHP ; b) 50 % DEHP ; c) 40 % DEHP ; d) 30 % DEHP : ▲ Sunflower, ● ethanol , ▪ acetic acid , ⁎ water.

- **GC Determination of DEHP Migration**

The effect of the initial concentration of DEHP on its specific migration in sunflower oil is illustrated by Figure 3. It is observed that the amount of migrating DEHP is related to its initial concentration in the PVC discs, to the temperature of migration testing and to the presence or absence of agitation. It is obvious that the mobility of the plasticizer molecules increased with increasing temperature and that the migration is favoured by agitation.

Figures 4 and 5 illustrate the effect of the nature of food simulant on the migration of DEHP at 45°C and 25°C with agitation, respectively. In all the cases, it is observed that the amounts of migrating DEHP are more pronounced in sunflower oil than in aqueous ethanol. This is due to the better solubility of DEHP in sunflower oil. Furthermore, for a same initial concentration of plasticizer, the amounts of migrating DEHP are higher at 45°C in the two food simulants. This is due to the fact that the mobility of the plasticizer molecules increased with increasing temperature.

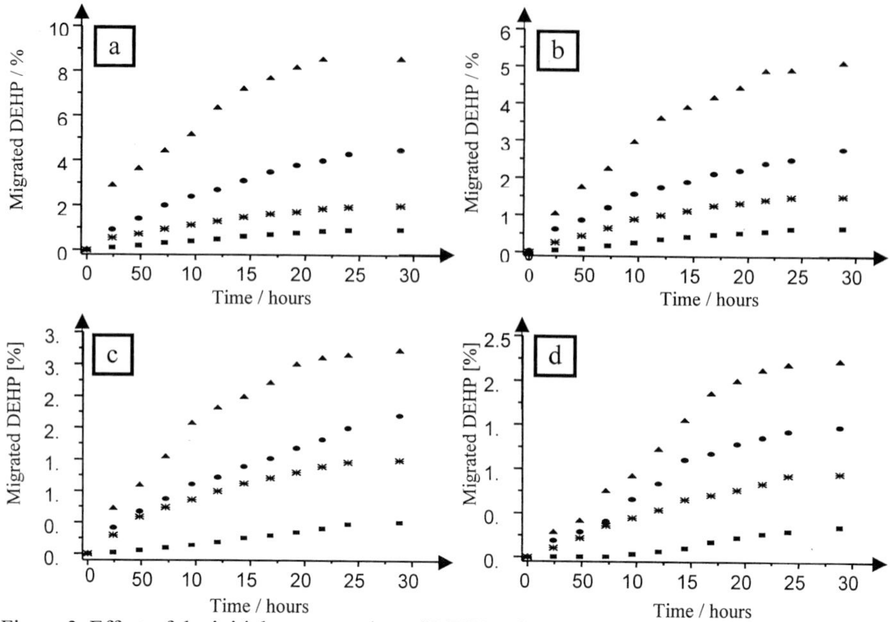

Figure 3. Effect of the initial concentration of DEHPon its specific migration in sunflower oil: a) 45 °C with agitation; b) 45 °C without agitation; c) 25 °C with agitation; d) 25 °C without agitation: ▲ 60% DEHP, ● 50% DEHP, * 40% DEHP, ▪ 30% DEHP.

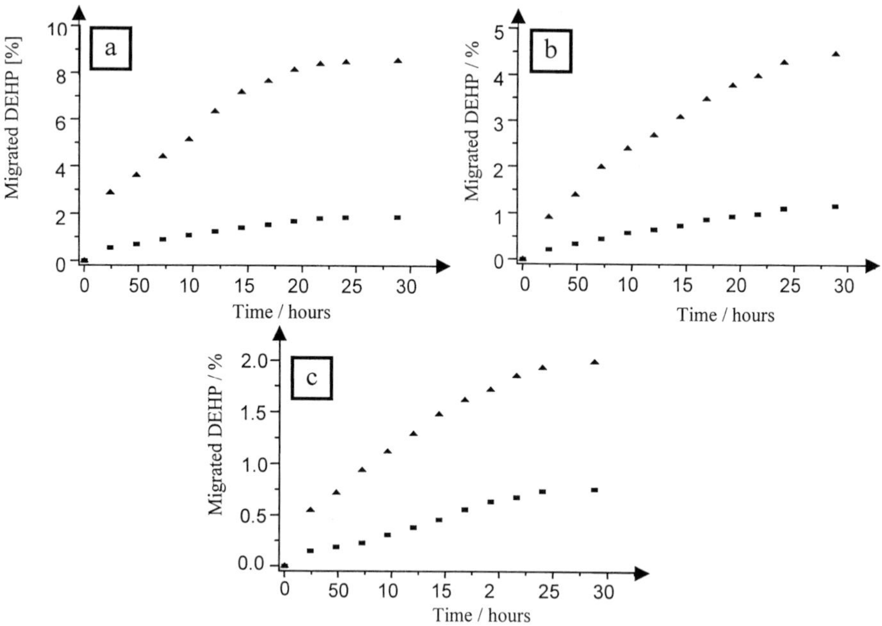

Figure 4. Effect of the nature of food simulant on the migration of DEHP at 45 °C with agitation : a) 60 % DEHP ; b) 50 % DEHP ; c) 40 % DEHP : ▲ sunflower oil, ▪ ethanol.

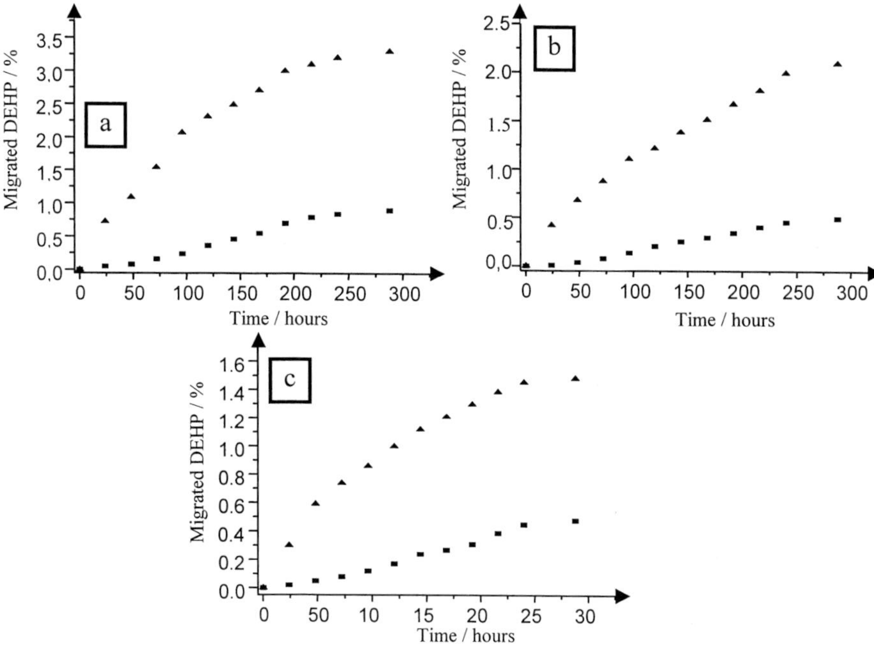

Figure 5. Effect of the nature of food simulant on the migration of DEHP at 25 °C with agitation : a) 60 % DEHP ; b) 50 % DEHP ; c) 40 % DEHP : ▴ sunflower oil, ▪ ethanol.

- **FTIR Investigation**

The direct analysis of the spectra of food simulants hasn't permit to show the migration of additives because the overlapping of characteristic bands and the very low amounts of migrated additives. On the other hand, the investigation of the PVC films spectra evidenced the DEHP migration as shown by Figure 6.

A progressive decrease until disparition of the plasticizer bands located at 1585 cm^{-1}, 1128 cm^{-1} and 742 cm^{-1}. was observed; Figure 7 illustrates the variation of their respective absorbances as a function of time.

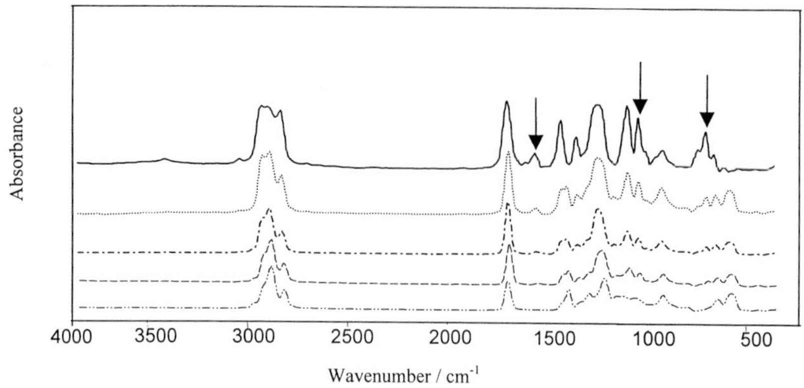

Figure 6. FTIR spectra of DEHP and PVC films obtained from the discs which were submitted to the migrating test with sunflower oil at 45°C with agitation. The initial content of DEHP in the PVC discs was 60 % : ——— DEHP, ... 0 day, -.-.-.. 1 day, – – – - 5 days, —..—.. 9 days.

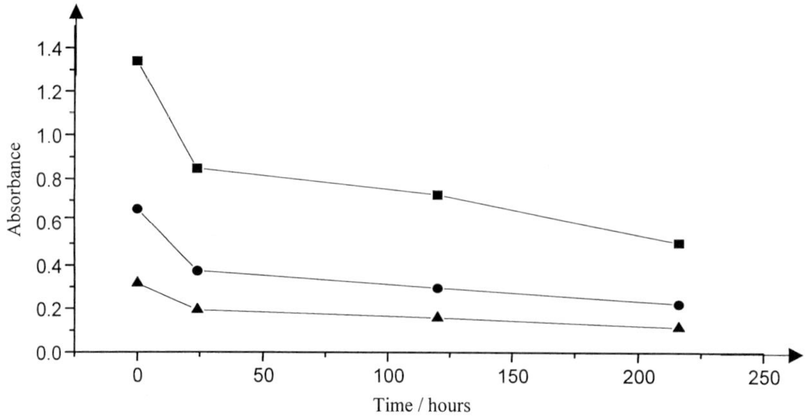

Figure 7. Variation of the absorbances of characteristic bands of DEHP as a function of time : —▲— 1585 cm^{-1}, —■— 1128 cm^{-1}, —●— 742 cm^{-1}.

- **DSC Analysis**

It is known that the presence of a plasticizer decreases the glass transition temperature (Tg) of PVC. For that purpose, the Tg's of virgin PVC and of four formulations containing 0, 10, 20 and 30 % of DEHP in addition to the other additives mentioned in Table 1 were determined. The results are given in Table 2 . It can be noted that the values of Tg decreased with increasing the amount of the plasticizer.

Table 2. Glass transition temperatures (Tg) for various formulations of PVC.

Sample	Tg °C
Virgin PVC	81.00
0 % DEHP	69.40
10 % DEHP	53.50
20 % DEHP	33.30
30 % DEHP	15.40

An eventual migration of DEHP will result in an increase of Tg. Hence, the Tg's of PVC discs, whose initial content of DEHP was 30 % and which were submitted to migrating test in sunflower oil at 45 °C with agitation, were determined. The results are given in Table 3. The Tg varied from an initial value of 15.40 °C to 48.50 °C after 10 days of contact with sunflower oil. This last value is close to the one obtained for a PVC containing 10 % DEHP and evidenced clearly the migration of the plasticizer.

Table 3. Tg's values after different times of contact with sunflower oil.

Sample	Tg °C
0 day	15.40
4 days	26.20
10 days	48.50

- **Atomic Absorption Spectrometry Investigation**

The amounts of metals contained in all the additives used as well as in the virgin sunflower oil were first determined. The results are given in Table 4.

Table 4. Amounts of metals in additives and sunflower oil.

Content (mg/l)	[Cu]	[Ca]	[Fe]	[Zn]	[Pb]	[Sn]	[Cd]	[Ni]
DEHP	*	8.5	*	*	*	*	2.6	*
Stabilizer	*	4.3	*	*	*	788.6	*	4.7
Loxiol G78	31.1	46.0	*	28.2	44.8	21.3	*	*
Loxiol GH4	18.0	21.6	3.0	6.2	21.1	*	4.1	*
Sunflower oil	6.0	*	4.2	2.1	*	*	3.5	1.1

*: undetected

The same metals were searched for in various samples of sunflower oil which were in contact with the PVC discs during the migrating test realized at 45 °C with agitation. The results are given in Table 5. It is observed that the concentrations of Cu, Ca, Zn, Pb and Sn increased with the time of contact. The presence of the last metal in relatively high amounts can be related to the migration of the stabilizer while the presence of the others can be related to the migration of the lubricants (Loxiol G78 and Loxiol GH4).

Table 5. Amounts of metals detected in sunflower oil as a function of time in the case of the migrating test carried out with the formulation whose initial content of plasticizer was 50 %.

Time	[Cu]	[Ca]	[Fe]	[Zn]	[Pb]	[Sn]	[Cd]	[Ni]
3 days	32.0	50.0	*	9.2	10.5	395.0	*	*
8 days	34.5	60.	*	12.6	15.5	465.0	*	*
10 days	39.0	90.3	4.6	16.0	22.5	485.0	*	*
12 days	44.5	97.5	5.4	18.1	32.0	650.5	3.8	*

*: undetected

Conclusion

In the preliminary study of the mass variation, the results showed that the rates of mass variations decreased highly in sunflower oil indicating that migration of additives occurred while for aqueous simulants (15 % v/v aqueous ethanol, 3 % w/v aqueous acetic acid and distilled water) the rates of mass variations increased very weakly and were practically the same indicating that penetration of aqueous simulant in the PVC discs occurred. Furthermore, the highest values of the rates of mass variations were obtained for the migration tests with agitation and for the formulations containing the highest amounts of plasticizer.

The GC analysis of DEHP migration showed that the migration of this plasticizer is related to its initial concentration, to the temperature of migration testing and to the agitation. The highest migration of DEHP occurred in sunflower oil where its solubility is the best. FTIR and DSC investigations confirmed the migration of DEHP in sunflower oil.

The atomic absorption spectrometry showed an increase of the concentrations of Cu, Ca, Zn, Pb and Sn in various samples of sunflower oil which were in contact with the PVC discs. The presence of the last metal can be related to the migration of the stabilizer while the presence of the others can be related to the migration of the lubricants.

All the results obtained with the various analytical methods used (GC, FTIR, DSC and AAS) showed that the migration of the additives present in the PVC formulations such as stabilizer, plasticizer and lubricants occurred. The phenomenon is influenced by the nature of food simulant, the temperature, the agitation and the initial concentration of the plasticizer.

[1] M. Reyne, "*Les plastiques dans l'emballage*," édition HERMES, Paris **1999**, p 28.
[2] K. Figge, *Fd cosmet. Toxicol.* **1972,**10, 815.
[3] S.G. Gilbert, *Journal of food Science* **1976,** 41, 955.
[4] L. Castle, A. J. Mercier, J. R. Startin , J. Gilbert, *food addit. Contam.* **1987**, 4, 399.
[5] M. Pizzoli, M. Scandola, CDROM, "*The polymeric materials Encyclopedia* ", CRC Press Inc, **1996**, p 1.
[6] E. Kondyli, P.G. Demertzis, M. G. Kontominas, *Food Chemistry* **1992**, 45, 163.
[7] S. S Chang, C. M. Guttman, I. C. Sanchez, L. E. Smith, "*Food and packaging interactions*", Edited by J. H. Hotchkiss, ACS symposium series 365, **1988**, p 106.

Macromol. Symp. ***180****, 203–215 (2002)*

Study of the Miscibility and the Thermal Degradation of PVC / PMMA Blends

Naïma Belhaneche - Bensemra[1], *Belkacem Belaabed*[2], *Abdelmalik Bedda*[1]

[1]: Laboratoire de recherche en sciences et techniques de l'Environnement, Ecole Nationale Polytechnique, BP 182, El-Harrach, Alger, Algérie (E-mail: nbelhaneche @yahoo.fr)

[2]: Ecole Supérieure des Techniciens de l'Aéronautique, Dar-El-Beida, Alger, Algérie

Summary: The aim of this paper is to study the miscibility and the thermal degradation of PVC/PMMA blends. For that purpose, blends of variable compositions from 0 to 100 wt % were prepared with and without plasticizer. Their physico-chemical characterization was carried out by differential scanning calorimetric analysis (DSC) and Fourier transform infrared spectroscopy (FTIR). Their thermal degradation under nitrogen at 185°C was studied and the HCl evolved from PVC was measured by the pH method. Degraded samples were characterized, after purification, by FTIR and UV- visible spectroscopy. The DSC analysis showed polymer miscibility up to 60 wt % of PMMA. This miscibility is due to a specific interaction of hydrogen bonding type between carbonyl groups (C=O) of PMMA and hydrogen from (CHCl) groups of PVC as evidenced by FTIR analysis. On the other hand, it was found that PMMA exerted a stabilizing effect on the thermal degradation of PVC by reducing the zip dehydrochlorination and by leading to the formation of short polyenes.

Introduction

In recent years blending of polymers has gained significant interest. It is the cheapest way of producing materials with improved properties without laborious development of new products. This field has become economically very important.[1]

Polyvinylchloride (PVC) is one of the most important and widely used thermoplastics due to its many valuable properties like low price, good processability, chemical resistance and low flammability. Its principal drawback, however, is low thermal stability at processing temperatures. In practice, this problem is overcome by the use of stabilizers.[2] Several polymers are mixed with PVC as polymer plasticizers or processing aids. Polymethacrylates, especially polymethylmethacrylate (PMMA), are used as processing aids for PVC.[3] The first study on PVC and PMMA was done by Schurer et al.[4], who concluded that PVC was partially miscible with atactic and syndiotactic PMMA but almost completely immiscible with isotactic PMMA. The PMMA and PVC blend is a well known system in which a hydrogen- bonding type of specific

 CCC 1022-1360/00/$ 17.50+.50/0

interaction involving the α - hydrogen of PVC and the carbonyl group of PMMA is expected.[5] Polymers and their blends are often processed in the melt, which makes the thermal stability of these materials of primary importance. Several studies have shown that interactions between a polymer or its degradation products and a second polymer in the same environment, as in a blend, can alter the decomposition of the individual polymers.[6-9] I.C. Mc Neill et al.[10, 11] studied the thermal degradation of PVC and PMMA mixtures and found that the monomer formed is methyl methacrylate (MMA) from early depolymerisation of PMMA initiated by Cl radicals propagating dehydrochlorination (DHC) of PVC at a much lower temperature than that of PMMA depolymerisation when heated alone. D. Braun et al.[12, 13] studying the thermal degradation of PVC with various polymethacrylates, have shown that longer n-alkylester groups and higher concentrations of the respective polymethylacrylate exhibit some stabilization of PVC, whereas smaller ester chains and low concentrations lead to destabilization. Furthermore, they could not detect an effect of the miscibility of the samples on the thermal degradation.

In the present work, the miscibility and the thermal degradation of PVC/PMMA blends are studied. For that purpose, blends of variable compositions from 0 to 100 wt % were prepared with and without plasticizer. Their miscibility was investigated by DSC and FTIR. Their thermal degradation under nitrogen at 185 °C was studied and the HCl evolved from PVC was measured by the pH method. Degraded samples were characterized, after purification, by FTIR and UV-visible.

Experimental

- **Materials**

Commercial grades of resins and additives listed in Table 1 were used as received. The K wert value of PVC is 67 according to the DIN 53-726 ; ρ (PVC) = 0.54 g/cm^3; ρ (PMMA) =1.18 g/cm^3.

Table 1. Compounds used in this study.

Compound	Source
PVC 4000M	ENIP – Skikda (ALGERIA)
PMMA	BASF (GERMANY)
Lead bibasic phosphite (heat stabilizer)	HENKEL (GERMANY)
Stearic acid (lubricant)	HENKEL (GERMANY)
Di (ethyl- 2 hexyl) phtalate (plasticizer)	BASF (GERMANY)

Rigid blends of PVC and PMMA of variable compositions from 0 to 100 wt % were realized. Melt mixing was performed at 185 °C on a two- roll mill, using 1 part per 100 parts PVC of lubricant and 4 parts per 100 parts PVC of stabilizer. The blends were then pressed in a hydraulic press at 185 °C for 3 min under a pressure of 150 MPa. They were water cooled under the same pressure for 2 min. The same blends of PVC and PMMA were prepared in the same conditions in the presence of 30 parts per 100 parts PVC of plasticizer (plasticized blends).

- **Glass Transitions**

The glass transitions temperatures were measured with a Perkin Elmer DSC- 7 apparatus at a heating rate of 20 °C / min.

- **Dehydrochlorination of the Blends**

The amount of liberated HCl in nitrogen (72 ml/min) was determined by using the continuous pH method at 185 ± 1°C.[14]

- **Purification of the Samples**

The samples were purified before and after dehydrochlorination by dissolution in tetrahydrofuran (THF), precipitation with distilled water and drying under vacuum at 40 °C during 72h. The purified samples were used for FTIR and UV-visible characterization.

- **FTIR Analysis**

The infrared spectra were recorded with a Philips type PU 9800 FTIR spectrophotometer at a resolution of 2 cm^{-1} using KBr pellets (1 wt %). For the deconvolution of the carbonyl bands a method using the software Grams 386 was used.

- **UV-Visible Analysis**

The UV-visible spectra were recorded with a SECOMAM ANTHELIE LIGHT UV-visible spectrophotometer. Solutions of the samples were prepared in distilled THF (1g/l) and analyzed immediately.

Results and Discussion

The glass transitions temperatures (Tg) of rigid and plasticized PVC/PMMA blends are reported in Tables 2 and 3, respectively. The glass transition is an important feature of a blend. For miscible systems, a single Tg is observed[5]. From the obtained results and according to this Tg criterion of miscibility, this polymer pair appears to be miscible up to 50 wt % PMMA. It is known that the presence of a plasticizer decreases the Tg value. In our case and for a same composition all the plasticized blends showed lower values of Tg in comparison with the rigid blends.

Table 2. Glass transition temperature of rigid PVC /PMMA blends.

wt % of PMMA	Glass transition temperature / ° C		
0		83.00	
10		84.00	
20		84.50	
30		85.00	
40		85.80	
50		96.00	
60	88.70		112.40
70	96.00		113.60
80	100.50		113.60
90	103.50		114.50
100		115.00	

Table 3. Glass transition temperature of plasticized PVC/PMMA blends.

wt % of PMMA	Glass transition temperature / ° C		
0		51.80	
10		57.00	
20		69.50	
30		75.00	
40		84.80	
50		97.00	
60	85.50		111.30
70	86.00		112.00
80	98.50		113.60
90	104.50		114.50
100		115.00	

The glass transition temperatures measured within the miscibility domain (up to 50 wt % PMMA) are reported as a function of blend composition in Figures 1 and 2 for both rigid and plasticized blends considered, respectively. A large positive deviation is observed indicating that the two polymers present a favorable specific interaction as reported by S. Shen et al.[15]

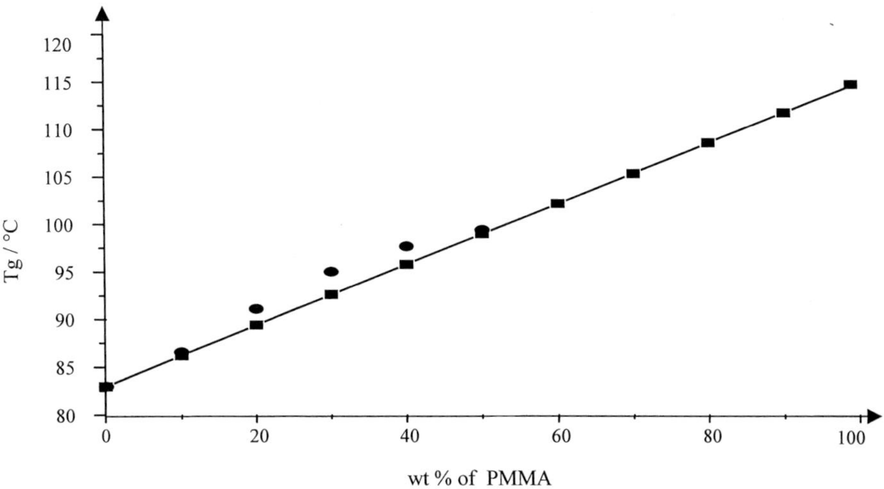

Figure 1. Glass transition temperature as a function of PMMA content in rigid PVC /PMMA blends. The line represents the behaviour based on the simple weight function $Tg = w_1Tg_1 + w_2 Tg_2$, where w_1 and w_2 are weight fractions of components 1 and 2, respectively ; ■ $=Tg = w_1Tg_1 + w_2 Tg_2$, ● = experimental Tg.

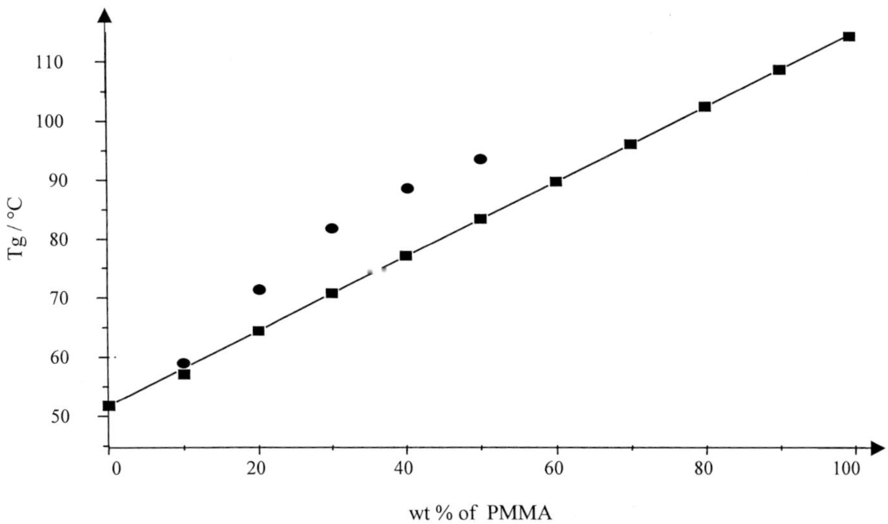

Figure 2. Glass transition temperature as a function of PMMA content in plasticized PVC /PMMA blends. The line represents the behaviour based on the simple weight function $Tg = w_1Tg_1 + w_2 Tg_2$, where w_1 and w_2 are weight fractions of components 1 and 2, respectively ; ■ $= Tg = w_1Tg_1 + w_2 Tg_2$, ● = experimental Tg.

FTIR analysis of the blends after purification showed a shift of the carbonyl band of PMMA to lower wavenumbers. The shift of the peak is about 3.5- 3.6 cm^{-1} within the domain of miscibility of the two polymers. This feature indicates that the miscibility of PVC/PMMA blends is due to a specific interaction of hydrogen bonding type between carbonyl groups (C= O) of PMMA and hydrogen from (CHCl) groups of PVC.

In the case of carbonyl stretching bands, the high mass of the oxygen atom and the great rigidity of the bond lead to a new band which permits not only the identification of associated and non-associated groups, but also their quantification.

The deconvolution of the carbonyl band within the miscibility domain showed two contributions one corresponding to non bonded carbonyl groups in the blend (1733- 1737 cm^{-1}) and one at lower wavenumbers corresponding to hydrogen bonded carbonyl groups (1720-1723 cm^{-1}) (Figures 3 and 4).).

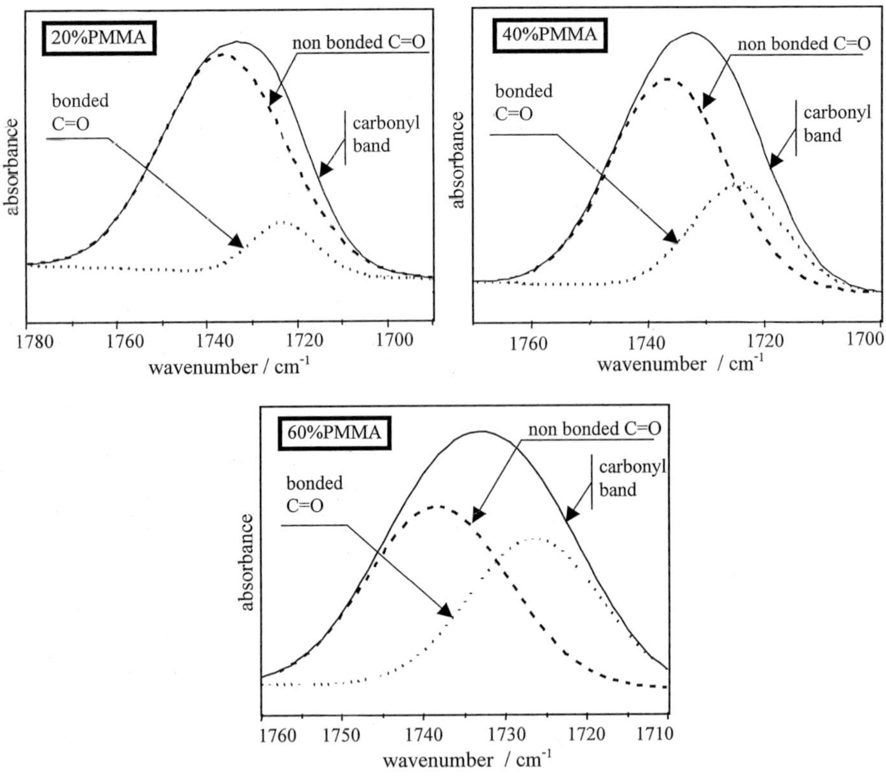

Figure 3. Deconvolution of the carbonyl band within the miscibility domain of rigid PVC/PMMA blends.

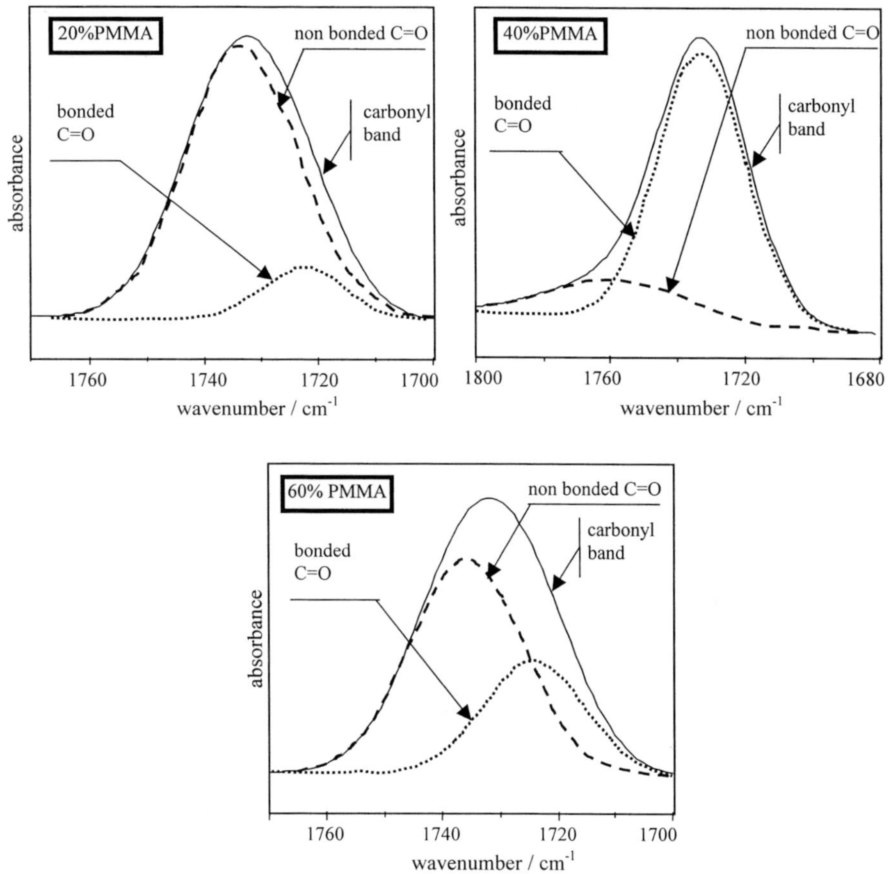

Figure 4. Deconvolution of the carbonyl band within the miscibility domain of plasticized PVC/PMMA blends.

The deconvolution of the carbonyl band outside the miscibility domain showed only contribution corresponding to non bonded carbonyl groups in the blend (Figures 5 and 6)

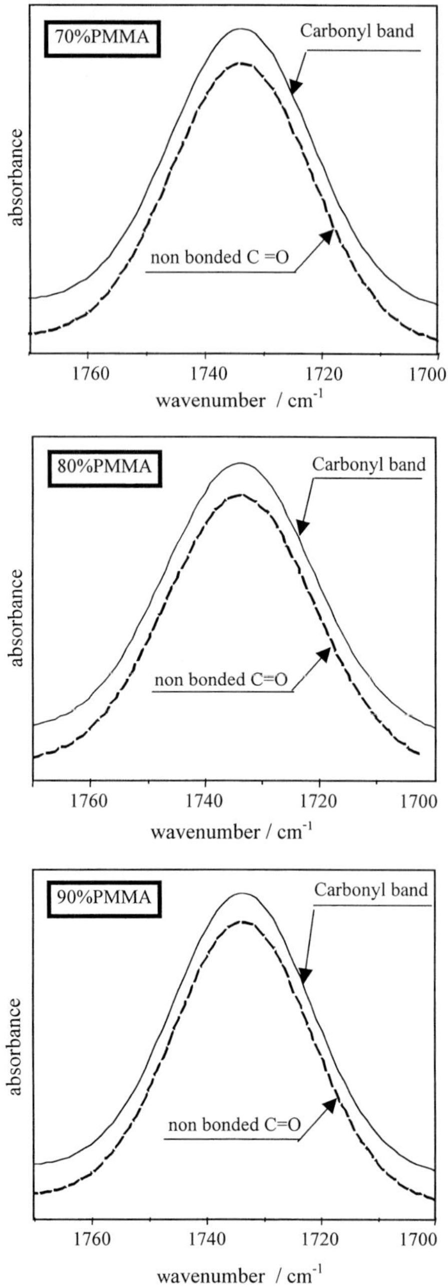

Figure 5. Déconvolution of the carbonyl band outside the miscibility domain of rigid PVC/PMMA blends.

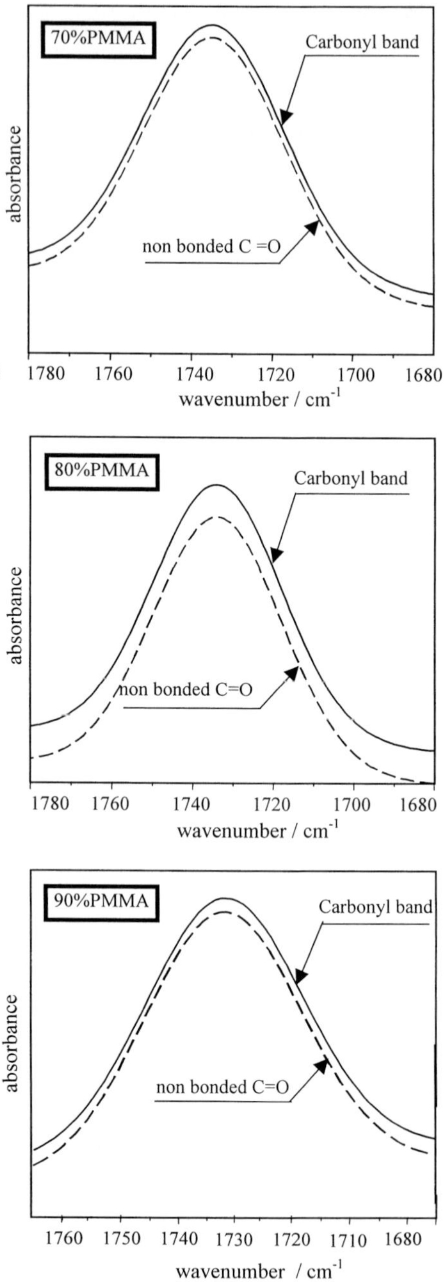

Figure 6. Déconvolution of the carbonyl band outside the miscibility domain of plasticized PVC/PMMA blends.

The thermal dehydrochlorination (DHC) of the rigid and plasticized PVC/ PMMA blends was measured at a constant temperature (185°C) in an inert atmosphere to exclude disturbing reactions of oxygen.

Figures 7 and 8 show the corresponding degradation curves.In all the blends PVC is more stable than PVC alone and the thermal stability increases with increasing the amount of PMMA in the blend.

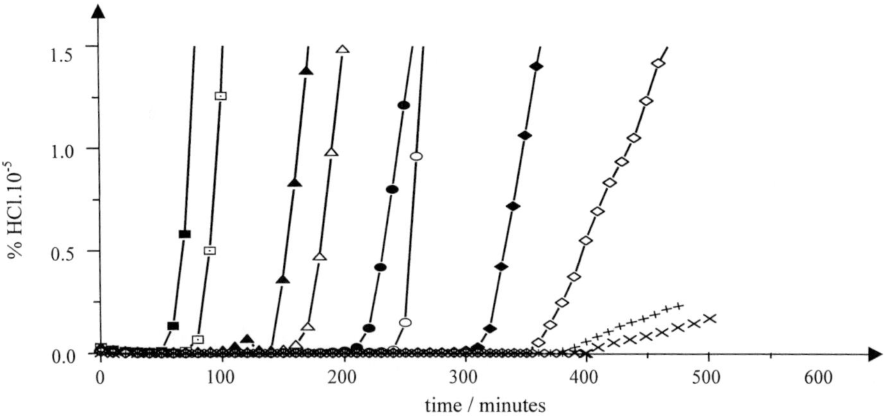

Figure 7. Dehydrochlorination curves of rigid PVC/PMMA blends at 185 °C in nitrogen; ■ = PVC, □ = 90/10, ▲ = 80/20, △ = 70/30, ○ = 60/40, ● = 50/50, ◆ = 40/60, ◇ = 30/70, + = 20/80, × = 10/90.

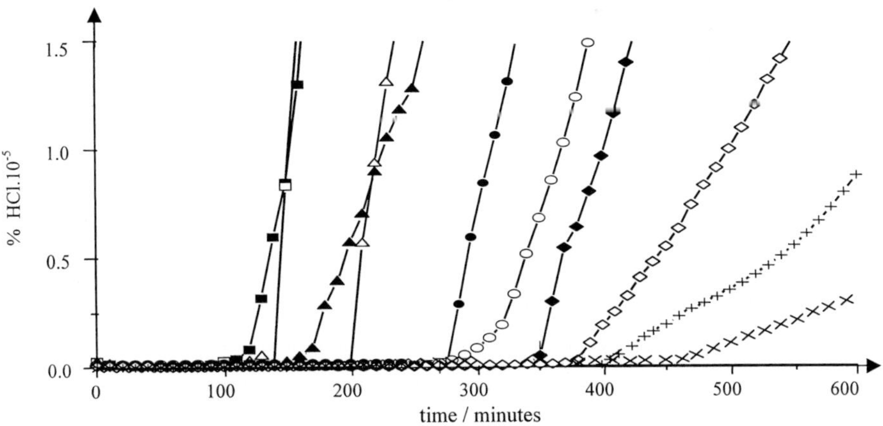

Figure 8. Dehydrochlorination curves of plasticized PVC/PMMA blends at 185 °C in nitrogen; ■ = PVC, □ = 90/10, ▲ = 80/20, △ = 70/30, ● = 60/40, ○ = 50/50, ◆ = 40/60, ◇ = 30/70, + = 20/80, × = 10/90.

Furthermore, for a same blend composition, plasticized mixtures showed longer induction times than rigid mixtures (Figure 9). The induction time is defined as the time before the accelerated DHC occurs. I. C. Mc Neill et al.[10,11] explained the best stability of PVC in presence of PMMA in terms of two processes occurring simultaneously during the degradation of the blend. The first is attack on PMMA by chlorine radicals produced during the DHC of the PVC; the second is the reaction between methacrylate ester groups and the hydrogen chloride. In our case, the presence of the heat stabilizer, which reacts with evolved HCl, gives a supplementary explanation of the observed features.

Furthermore, the linear variation of the induction time as a function of the amount of PMMA for both rigid and plasticized mixtures indicates the absence of a direct effect of miscibility on the thermal degradation of the blends.

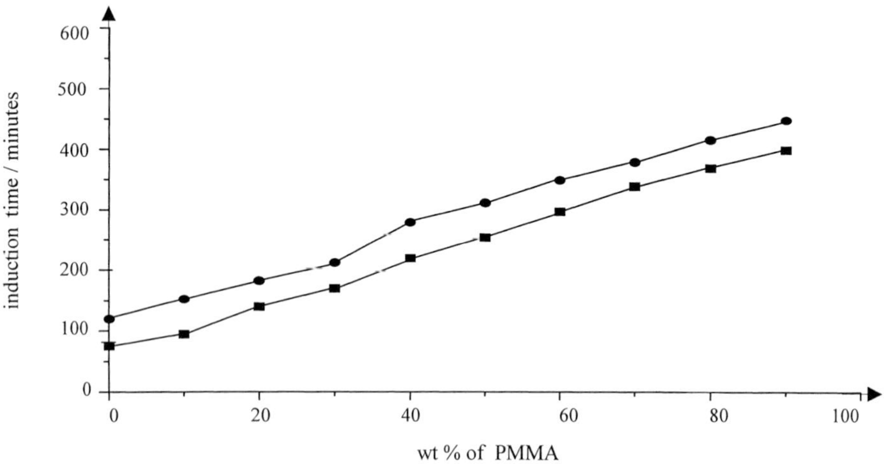

Figure 9. Variation of the induction time as a function of the amount of PMMA ; ■ = rigid blends , ● = plasticized blends.

An example of FTIR spectra of degraded samples after purification is shown in Figure 10. The appearance or disappearance of characteristic bands were not observed, but a noticeable decrease of the intensity of all the regions of the spectra was observed showing that DHC of PVC and depolymerization of PMMA have occured.

The degraded samples of PVC/PMMA blends were investigated after purification with UV-visible spectroscopy, too (Figures 11 and 12). All the spectra present the same aspect. They show clearly that the polyenes formed in the PVC in such mixtures absorb in the UV region at $\lambda < 300$ nm which corresponds to short polyenes having a number of conjugated double bonds $n \leq 4$.[16]

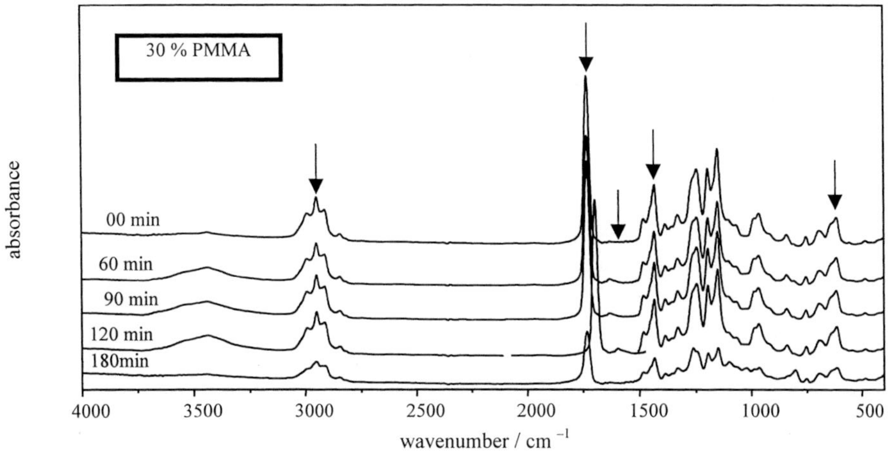

Figure 10. FTIR spectra of the plasticized blend PVC/PMMA (70/30) at various times of degradation at 185 °C.

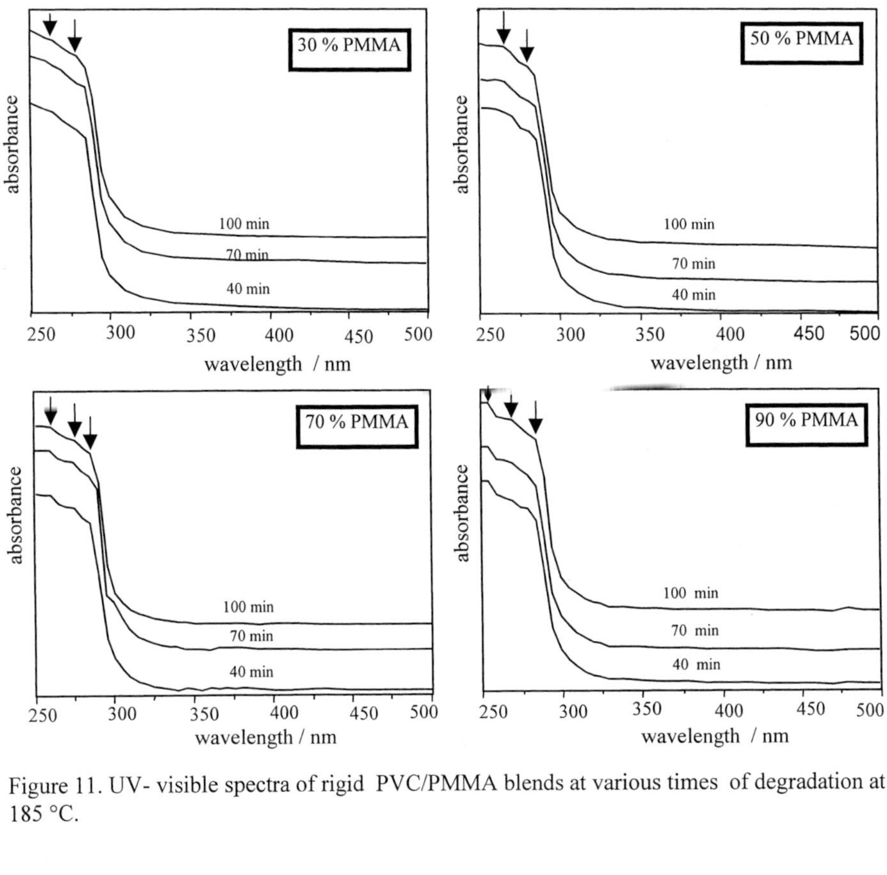

Figure 11. UV- visible spectra of rigid PVC/PMMA blends at various times of degradation at 185 °C.

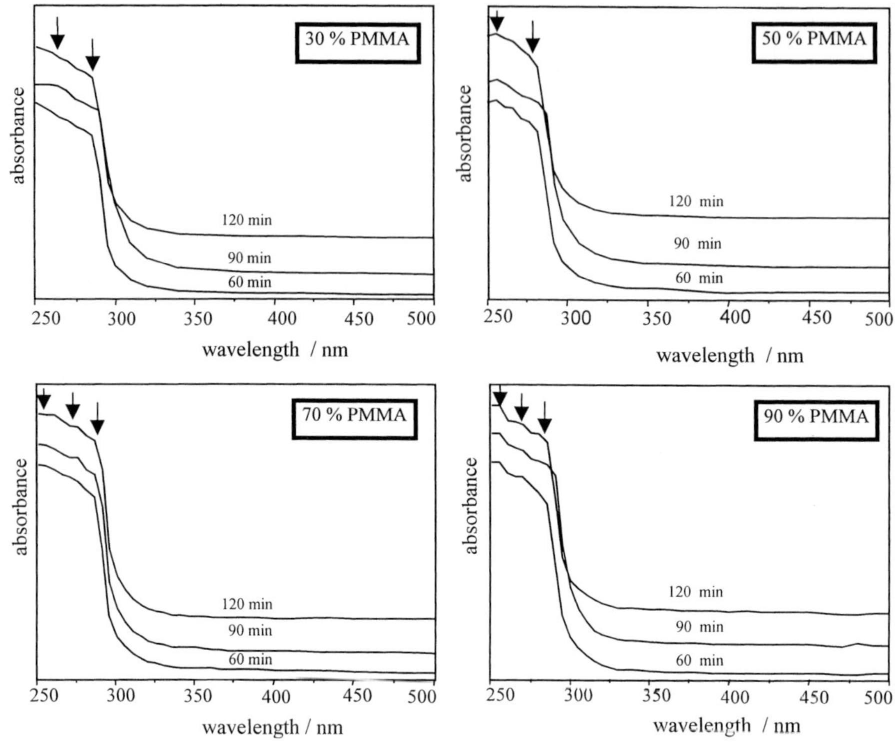

Figure 12. UV-visible spectra of plasticized PVC/PMMA blends at various times of degradation at 185 °C.

Conclusion

The DSC analysis of the rigid and plasticized PVC/PMMA blends showed polymer miscibility up to 50 wt % of PMMA. This miscibility is due to a specific interaction of hydrogen bonding type between carbonyl groups (C= O) of PMMA and hydrogen from (CHCl) groups of PVC as evidenced by FTIR analysis.

On the other hand the results showed that PMMA exerted a stabilizing effect on the thermal degradation of PVC by reducing the zip dehydrochlorination and by leading to the formation of short polyenes. No influence of the miscibility of the two polymers was detected on their thermal degradation.

[1] L.A. Utracki, A. Ajji, M. M. Dumoulin, Alloys and blends, CDROM, " *The Polymeric Materials Encyclopedia*", CRC Press Inc. **1996**, p. 1.
[2] D. Ausseur, Le polychlorure de vinyle, " *Techniques de l'Ingénieur*", A3326, **1990**, p 1.

[3] M. Reyne, ''*Les plastiques : polymères, transformations et applications*'', Edition HERMES, Paris **1992**, p. 96.
[4] J.W. Shurer, A. de Boer, G. Challa, *Polymer* **1975**, 16, 201.
[5] H.Jager, E. J. Vorenkamp, G. Challa, *Polymer Commun.* **1983**, 24, 290.
[6] A. Kaminska, H. Kaczmarek, *J. Thermal Anal.* **1984**, 29, 1017.
[7] L . Goulet, R. E. Prud'homme, *Eur. Polym. J.* **1986**, 22, 529.
[8] E. J. Moskala, D. W. Lee, *Polym. Degr. Stab.* **1989**, 25, 11.
[9] K. Kuzelova, Z. Vymazal, *Eur. Polym. J.***1999**, 35, 361.
[10] I. C. Mc Neill, D. Neill, *Eur. Polym. J.***1970**, 6, 143.
[11] I. C. Mc Neill, D. Neill, *Eur. Polym.* J.**1970**,6, 569.
[12] D. Braun, B. Böhringer, W. Knoll, N. Eidam,W. Mao, Die Angew. Makromol. Chem. **1990**, 181, 23.
[13] D. Braun, B. Böhringer, N. Eidam, M. Fisher, S. Kömmerling, Die Angew. *Makromol. Chem.* **1994**, 216, 1.
[14] ISO 182-2, First edition, 1990-12- 15.
[15] S. Shen, J.M. Torkelson, *Macromolecules* **1992**, 25, 721.
[16] K.B. Abbas, R. Laurence, *J. Polym. Sci. : polym.Chem.* Ed. **1975**, 13, 889.

Macromol. Symp. ***180,*** *217–221 (2002)*

Synthesis and Characterization of Poly(3-alkylthiophenes) with NLO Chromophoric Groups in Side Chains

C. Della Casa, A. Fraleoni, P. Costa Bizzarri, M. Lanzi and L. Paganin*

Dipartimento di Chimica Industriale e dei Materiali, Università di Bologna, Viale Risorgimento 4, 40133 Bologna, Italy

Summary: Polythiophene copolymers containing alkyl side chains of different length and partially functionalized with chromophoric groups have been synthesized by $FeCl_3$ oxidative coupling of 3-alkylthiophenes and functionalized 3-alkylthiophenes. Composition, molecular weigth and configuration of the soluble fraction of the copolymers have been investigated and some structure-property correlations have been evidenced.

Introduction

Over the last decade, organic materials exhibiting non-linear optical (NLO) properties have been the subject of intensive investigations for their possible technological applications such as data storage and data transmission. In particular, polymers containing NLO active chromophores seem to be very promising for these purposes.[1,2] A first approach is to consider a polymer as an inert system in which to dissolve NLO active molecules (guest-host system). However such a dispersion of molecules presents many problems as, for instance, their usually low solubility within the polymer matrix which limits the magnitude of the NLO response. In order to overcome these problems the chromophores can be incorporated into the main or side chain of the polymer so as to enhance the chromophore concentration and consequently the non-linear activity.

In the present paper the chromophore incorporation method was tested on 3-alkyl substituted polythiophenes which are characterized by good thermal and chemical stability. A chromophore was grafted directly to the side chains of 3-alkylthiophenes and the functionalized monomers were then copolymerised with 3-alkylthiophenes as plastifying comonomers to improve solubility. New processable polythiophene-based materials functionalized with NLO chromophoric units were thus obtained.

 CCC 1022-1360/00/$ 17.50+.50/0

Experimental

All chemical reagents including 3-hexylthiophene (3-HT) and 3-dodecylthiophene (3-DDT) were purchased from Aldrich.

The functionalized monomers **1a** and **1b** (Figure 1) were synthesized by the corresponding ω-bromoalkylthiophenes[3,4] with the azo dye 4-hydroxy-4'-nitroazobenzene[5] in a polar solvent and in the presence of potassium carbonate. The yield of monomer **1a** can be improved by using the *p*-toluenesulfonate[6] derivative instead of the bromo derivative.

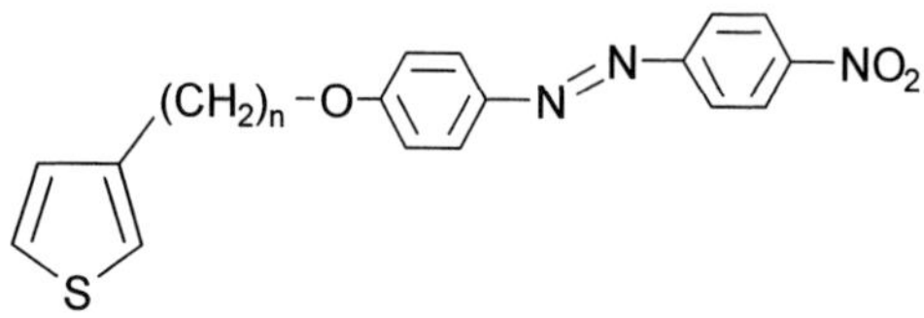

1a n = 2

1b n = 6

Figure 1. Structures of functionalized monomers.

Mixtures of 3-HT and **1a,** 3-HT and **1b**, and 3-DDT and **1b**, in a 1:1 mole ratio, were copolymerized to give copolymers **2a**, **2b**, and **2c** (Figure 2), respectively.

Copolymerization (procedure A) was carried out over a 1 hr period by initially adding a solution of $FeCl_3$ in nitromethane to a solution of the comonomers in carbon tetrachloride.[7] The final concentrations of the mixture of the two comonomers and $FeCl_3$ were fixed to 0.06 M and 0.24 M, respectively. The reaction mixture was treated with acidic methanol to precipitate the polymer and to change it to the neutral state from the form complexed with $FeCl_3$. The polymer was extracted in a Soxhlet apparatus with methanol in order to remove residual oxidant and low molecular weight materials and then with chloroform to obtain a soluble fraction. In the preparation of polymer **2a** a different procedure (B) was also experimented to obtain a higher yield of soluble fraction, consisting in the initial addition of pure carbon tetrachloride to a solution of $FeCl_3$ and comonomers in nitromethane.

The molecular weights of the polymers were determined by gel permeation chromatography (GPC) in THF using polystyrene standards as references. Polymer composition and microstructure were assessed by ^{1}H-NMR spectroscopy in $CDCl_3$

solutions. Glass transition temperatures (T_g) were measured by DSC at a heating rate of 20°C min^{-1}.

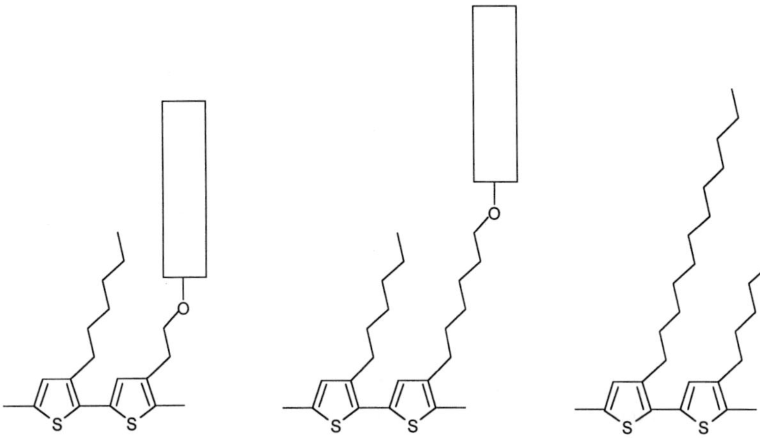

2a 2b 2c

Figure 2. Structures of functionalized copolymers with different types of side chains.

Results and discussion

The azo dye 4-hydroxy-4'-nitroazobenzene was choosen as chromophore because it was easily grafted to the alkyl side chain of the thiophene ring. Nucleophilic substitution reactions with the appropriate thiophene intermediates led to the monomers **1a** and **1b** with a molecular hyperpolarizability, a second order NLO parameter, $\beta = 127 \times 10^{-30}$ and 144×10^{-30} esu (measured at $\lambda = 1.06$ μm), respectively.[8]

Copolymerizations of the two monomers with procedure A (see Experimental) gave good polymer yields, that is 61 % (**2a**), 79 % (**2b**), and 61 % (**2c**), with fractions soluble in solvents such as chloroform (Table 1), from which films could be cast onto substrates. A very low amount of soluble polymer **2a** was however obtained with this procedure (see Table 1). A very substantial increase of the soluble fraction from 4 to 33% for **2a** was achieved by adopting the alternative procedure B for reagent addition. In this case the overall polymer yield is about half that obtained by method A while almost all the polymer is soluble. It is worth noting that polymer **2a** prepared according to either procedure exhibits similar composition, number average molecular weight (M_n) and polydispersity index (M_w/M_n).

Table 1. Yields and some characteristic data for the soluble fractions of the copolymers.

Copolymer (procedure)[a)]	Soluble fraction (%)[b)]	Monomer **1a** or **b** (mol%)[c)]	$M_n \times 10^{-3}$ (g/mol)	M_w/M_n	Regioregularity (HT%)[d)]
2a (B)	33	31	5.5	2.5	61
2a (A)	4	30	6.0	2.7	64
2b (A)	23	37	6.6	3.1	69
2c (A)	16	28	11.0	2.5	75

a) See experimental part.
b) Referred to the initial weight of the comonomers.
c) In the copolymer.
d) HT = head-to-tail coupling.

The length of the flexible side chains seems an important parameter not only for polymer solubility but also for its microstructure. The latter was assessed by ^{1}H-NMR according to regular head-to-tail (HT) coupling content calculated on the basis of the intensity of the signals relative to methylene protons in α to the thiophene ring.[7,9] As can be seen in Table 1, the HT configuration content increases as the length of the alkyl chain increases in either one or the other comonomer. The lowest value is found for polymer **2a,** regardless of the polymerization method adopted, whether A or B.

A reverse trend was found for the glass transition temperature in the series of the polymers obtained with the procedure A; T_g, which is 50°C for **2a**, 40°C for **2b**, and -17°C for **2c**, decreases as the length of the alkyl chain increases.

A characteristic common to all the synthesized copolymers is the difference between comonomers feed ratio (1:1 mole ratio) and final polymer composition. The amount of chromophoric monomer in the soluble fractions is at best 37 mole% (Table 1) instead of the expected 50%. In comparison, the insoluble part of the polymers is always enriched in this monomer, as can be seen from the relative intensity of the bands in the FT-IR spectra of both soluble and insoluble parts.

Conclusions

In conclusion, a copolymerization route for obtaining new soluble poly(3-alkylthiophene)s functionalized with NLO chromophoric groups was successfully found. The polymerization procedure appears to be of primary importance for achieving a good amount of soluble polymer. Structure-property correlations for the new polymers were investigated and the importance of side chain length clearly emerged.

Acknowledgements

This work was supported by the Ministero dell'Università e della Ricerca Scientifica e Tecnologica and by the Università di Bologna (funds for selected research topics), Italy.

[1] D. M. Burland, R. D. Miller, C. A. Walsh, *Chem. Rev.* **1994**, *94*, 31.
[2] C. Samyn, T. Verbiest, A. Persoons, *Macromol. Rapid Commun.* **2000**, *21*, 1.
[3] E. Baciocchi, V. Mancini, P. Perucci, *J. Chem. Soc. Perkin* II **1975**, 821.
[4] P. Bäuerle, F. Würthner, S. Heid, *Angew. Chem. Int. Ed. Engl.* **1990**, *29*, 419.
[5] K.-H. Bantel, H. Musso, *Chem. Ber.* **1969**, *102*, 696.
[6] C. Della Casa, A. Fraleoni, P. Costa Bizzarri, M. Lanzi, *Synth. Met.*, in press.
[7] P. Costa Bizzarri, F. Andreani, C. Della Casa, M. Lanzi, E. Salatelli, *Synth. Met.* **1995**, *75*, 141.
[8] R. Pizzoferrato, F. Sarcinelli, M. Angeloni, M. Casalboni, F. Bertinelli, P. Costa Bizzarri, C. Della Casa, M. Lanzi, *Chem. Phys. Lett.* **2001**, *343*, 205.
[9] M. Sato, H. Morii, *Macromolecules* **1991**, *24*, 1196.

Textiles Thermosetting by Microwaves

Maria Cristina D'Arrigo[a], *Bonaventura Focher*[a]*, *Gian Carlo Pellacani*[a], *Cesare Cosentino*[b] *and Giangiacomo Torri*[b]

[a] Department of Chemistry - University of Modena and Reggio Emilia – Modena - Italy
[b] Institute of Chemistry and Biochemistry "G. Ronzoni" - Milan - Italy

Summary: Thermosetting is an important part of the finishing of thermoplastic poly(ethylene terephthalate) (PET) fabrics and garments that confers stability in dimensions and shape as well as appropriate hand to the final product. Conventional thermosetting methods for PET include hot air and steaming treatments.
In the present work we used solid state NMR as well as DSC methods in order to investigate any differences in the behaviour of PET chips when annealed with either a conventional or a microwave technique.

Introduction

The everyday increasing interest in microwave applications in the industrial sector lies in the possibilities to perform very fast manufacturing cycles, to reduce production costs, to decrease power consumption, to improve the quality of the existing products, and to increase the efficiency and productivity of the processes. Most of the benefits deriving from microwave applications, however, become evident only after a careful laboratory experimental work, and

* Corresponding author: Facoltà di Ingegneria dei Materiali e dell'Ambiente - Università di Modena e Reggio Emilia - via Vignolese, 905 - 41100 Modena, Italy.
E-mail address: bfocher@unimo.it

 CCC 1022-1360/00/$ 17.50+.50/0

preliminary economical cost-benefits analysis is always full of uncertainties, being difficult the extrapolation of experimental data to the larger industrial scale. The advantages of a well designed microwave system, however, are still numerous and appealing; in particular:

- uniform and homogenous heating
- temperature profile inversion
- materials treatment at surface temperature lower than conventional
- high power densities, localised and in short times

Thermosetting is an important part of the finishing of poly(ethylene terephthalate) (PET) fabrics or garments to impart dimensional stability, shape, and appropriate hand to the product during using. Conventional methods for setting dimensional stability include either hot air or steaming treatments.

The aim of the present work is to verify whether microwave technology can be applied to thermoplastic PET thermosetting and to evaluate the quality of the obtained products with this kind of technique that leads to undoubted economical advantages deriving from the rapid heating rate.

Experimental

Sample preparation

The PET under study is and industrial product obtained in chips from Montefibre SpA. It shows to be not highly crystalline, as evidenced both from X ray diffractometry (not shown), solid state NMR and DSC analysis.

The PET chips as received were subjected to different annealing treatments at temperatures from 90° to 250° for 15-60 min, in both conventional air oven and microwave furnace (a multimode cavity CEM MAS 7000 Incinerator, 2.45 GHz, 950). Temperature of the samples in the microwave treatments was determined by a thermocouple immediately after the turning off of the microwave irradiation.

The amorphous PET was prepared by melting PET chips in DSC analyser and quenched immediately in liquid nitrogen.

Methods

Differential scanning calorimetry experiments

DSC measurements were carried out in a Perkin Elmer DSC-4 calorimeter. Samples were heated from 30° to 300°C with a heating rate of 10 °C/min under a nitrogen atmosphere.

Solid state NMR studies

The CP-MAS ^{13}C NMR spectra at 75.4 MHz were recorded at room temperature with a Bruker ASX-300 spectrometer, using a 7mm ZrO_2 rotor. The cross polarisation (CP) contact time was 1,2 ms while the repetition time and the ^{1}H 90° pulse were from 4 to 20 s and 3.7 μs, respectively. The full width at half-height of the reference glycine was 27 Hz.

Chemical shifts were measured with respect to glycine as external reference set at 42 ppm from silyltrimethylpropionic acid sodium salt (TSP). For each spectrum 128-500 transients were accumulated with 1K data points. The spinning rate was 4.0 KHz, and the sweep width 360 ppm. The spectra were manually phased.

^{13}C T_1 relaxation times were measured by the Torchia sequence using CP T_1 experiments[1] where the magnetisation associated with each carbon atom is reduced to zero during longitudinal relaxation.

^{1}H spin-lattice relaxation measurements ($T_{1\rho}$) were used to quantify spin-diffusion via ^{13}C nucleous observation. The measurements were carried out under high resolution conditions to identify single phases where magnetisation propagation occurred.[2] The sequences have been described by Voekel.[3] Data were elaborated with both mono- and bi-exponential function, using xwin-NMR software by Bruker.

Results

Differential scanning calorimetry experiments

In order to focus the optimal conditions for any textiles thermosetting treatment it is necessary to know the morphology of the material on 1-50 nm scale, a region which can strongly influence the mechanical properties of the material.

In order to determine macroscopic phase transitions of PET after conventional or microwave annealing preliminary DSC measurements were performed. It is known that isothermal crystallization of isotropic PET samples is mainly controlled by the annealing temperature. [4,5]

Thermograms of the quenched PET and of the as received material are reported in Figure 1.

The variation of the base line before the 100°C is a clear image of the glass transition (Tg) temperature typical of non ordered areas, while the exothermic peak around 130°C is explainable through the crystallisation of the fold surface layers. Comparing the ΔH values of the quenched PET and of the as received one it is evident that the industrial drying stage in vacuum at 105°C induces a partial reorganisation of the material.

The endothermic peak around 230°C is a clear image of melting as the transaction typical for the crystalline regions.

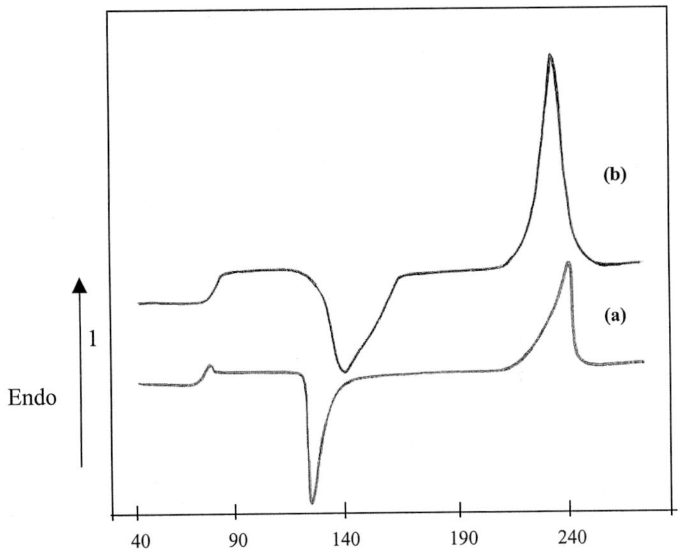

Figure 1 - DSC scanning of the as received (a) and quenched (b) PET samples.

The endothermic peak of cold crystallisation is not clearly visible any more already in the samples air annealed at 120°C (Figure 2), suggesting that when the system is brought towards a better mobility of the chains (T>Tg) the system has the tendency to reorganise in an ordered way when cooling down.

In the sample microwave annealed at 140 °C for 60 min the endothermic peak disappears faster than in the conventionally annealed sample indicating a prompter response of the material to thermal power. No differences in terms of ΔH_f (45-48 J/g) between air and microwave annealed samples were observed.

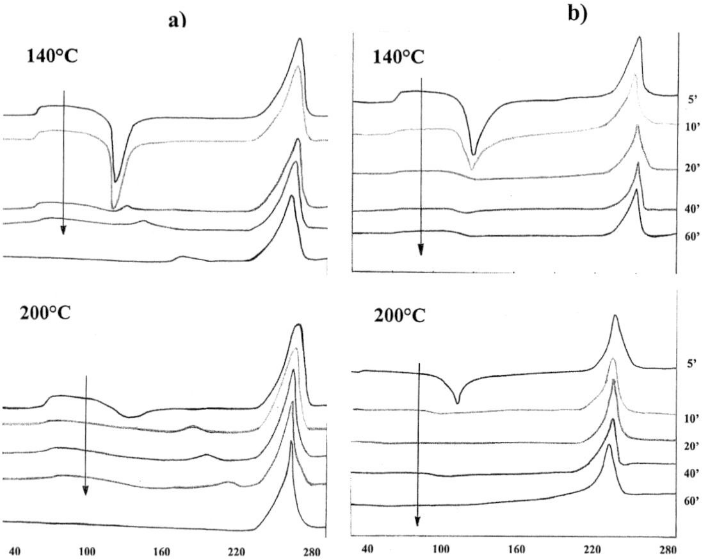

Figure 2 - DSC scans of the air (a) and microwave (b) annealed (140°-200°C) PET samples.

On the contrary by passing from 140° to 200°C a higher value of ΔH_f (80 J/g) was observed for the sample annealed in air in comparison to the sample heated in microwave furnace (52 J/g) suggesting the presence of a bigger and/or more perfect crystals in the first case.

Thus the DSC data evidence a faster structural change of PET samples annealed by microwaves when the temperature is lower than 160°C but at higher temperature the conventional heating is more effective. The effects of the two types of annealing are comparable beyond 180°C.

For many samples an endothermic peak called "mobile peak"[6] appears whose position depends strongly on the annealing temperature and is generally located a couple of degrees higher that the annealing temperature and it goes to higher temperatures with increasing the treatment time.

The origin of the peak seems to be attributed to the formation of ordered small areas located in the fold surfaces layers of the crystalline blocks. The mobile peak appears always for both the annealing treatments performed.

^{13}C CPMAS NMR spectra

Solid State NMR Spectroscopy differently from other techniques as X-Rays diffractometry and vibrational spectrometry is well indicated both for studying local structures, in particular different conformational situations, as well as for defining polymer dynamics. In fact, through anisotropic nuclear-spin interaction and the resulting angle dependent NMR frequencies, the nuclei can act as local probes of segmental orientations. [7,8]

The CP MAS NMR spectrum of the PET samples show (Figure 3) three groups of signals typical of the material composition, centered in the region of 162, 130 and 61 ppm. In the as received sample the signals are quite broad suggesting a dominant unordered structure contribution.

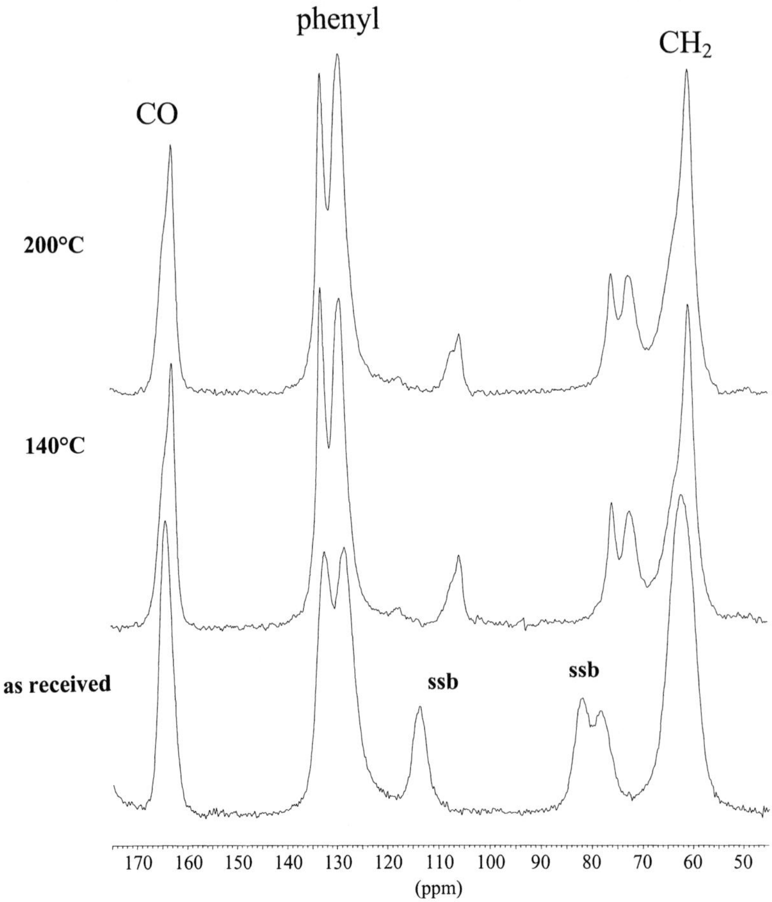

Figure 3 - CP MAS spectra of PET samples as received and at various air annealing temperatures.

Sample annealing both in air and in a microwave oven above glass transition temperature (Tg) induces a segment mobility of PET that brings to the formation of structures that seem more ordered, as evidenced from the narrowing of the signal and change in chemical shifts of the signals.

In fact, when passing from as received sample to the air annealed at 140° and 200°C for 60 min, a progressive reduction of the broadening of the signal of the samples is observed together with the co-existence of two components in the signals of the C=O (162.5 and 160 ppm) and CH_2 (60.5 and 60 ppm) groups.

Microwave annealing of the samples in the same experimental conditions of the conventional treatment brings to similar results, but less effective In fact, it is peculiar to point out the presence of two components in the C=O signal already for the sample treated at 140°C, but a slower kinetic in the structural change of the two termosetting treatments (Figure 4).

Even thought we do not expect large differences in the chemical shifts between the gauche and trans conformations of PET because the interactions with the substitutents in a γ position are small, nevertheless they are sufficient for making a distinction between the two PET conformations, the reduced mobility of the chain and the reduction in the broadening of the signals.

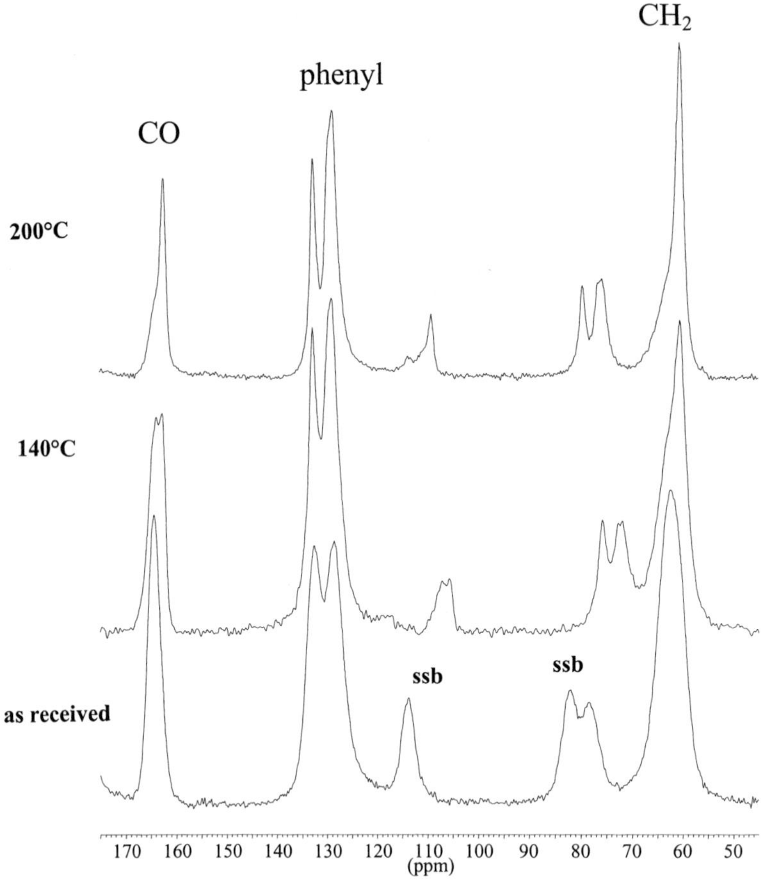

Figure 4- [13]CP MAS spectra of PET samples as received and at 140°-200°C microwave annealing temperatures.

When using the deconvolution of the signal at 60.5 ppm and integrating the two components it has been possible to determine the percentage of the trans and gauche conformation.[9,10] As an example, in Figure 5 is reported the CH_2 signal deconvolution of the air annealed (160°C) PET sample and its reconstructed profile.

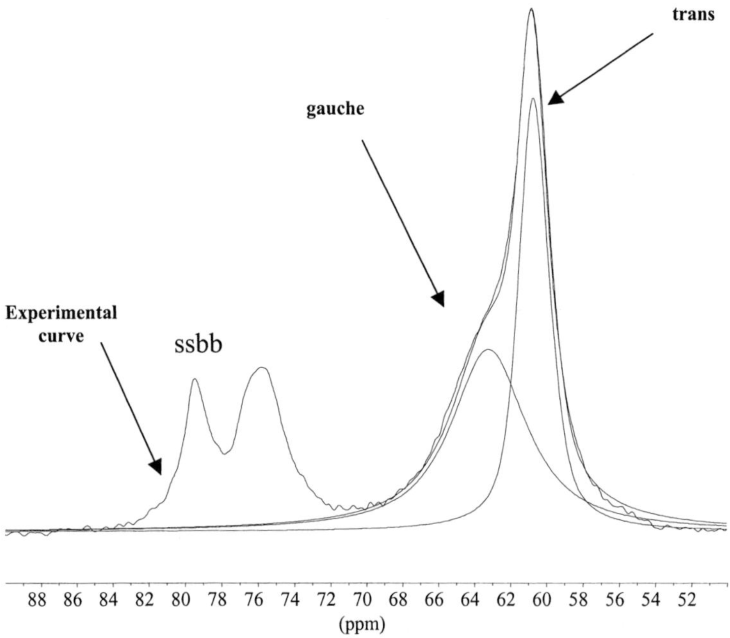

Figure 5 – Deconvolution of the methylen carbon CP-MAS resonance peak of PET (Gaussian and Lorentzian combination). The fitted curve is practically overlapping the experimental one.

In table 1 are reported the percentage of trans and gauche conformers obtained by deconvolution procedure for all samples.

Table 1 - Conformer distribution of methylen carbons.

sample	Air annealing temperature (60 min)	Trans (%)	Gauche (%)
PET	as received	3.5	96.5
	140°C	39.5	60.5
	160°C	37.0	63.0
	180°C	70.0	30.0
	200°C	>75	<25

sample	Microwave annealing temperature (60 min)	Trans (%)	Gauche (%)
PET	as received	3.5	96.5
	140°C	47.0	53.0
	160°C	43.0	57.0
	180°C	61.5	38.5
	200°C	>60	<40

The values of the trans conformation are stabilised after the treatment at 180°C.

The Magic Angle Spinning spectra (MAS) are sufficiently to point out the most mobile components of the system (Figure 6).[11]

The resonance of various groups, even though the MAS technique induces the loses of the more rigid components of the system, becomes narrower with increasing the temperature of treatment up to 160°C. Therefore, in spite a considerable number of scans, may be worthwhile to notice the presence of a limited amount of mobile parts in the as received sample that remains present till reaching the temperature of 160°C in the treatment.

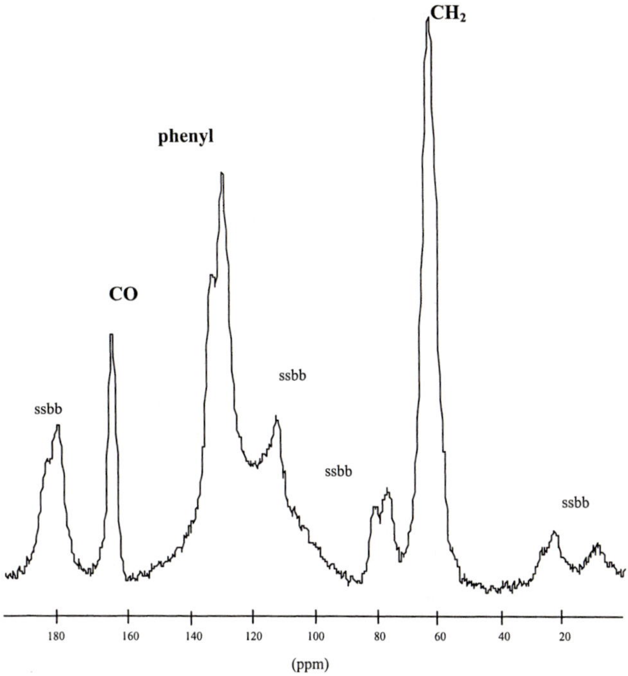

Figure 6 - Solid state NMR MAS spectra of PET.

Obviously in the quenched PET the mobile parts are much more numerous, indicating differently from what is shown in the XRD spectrum a significant crystallisation kinetic also in simple drying conditions for the material. This stresses out the importance of an NMR analysis to give direct information on local level without involving any result extrapolation.

Spin lattice relaxation times T_1 and $T_{1\rho}$

In order to study in detail the structure of the annealed PET materials we performed experiments of spin lattice relaxation times T_1 (Torchia) of the samples (Table 2).

Table 2- a) Single fitting of T_1-Torchia values (s)

PET sample		as received	140 °C	200 °C
ppm	162.5	14.3	63.9	158.1
	132.7	13.7	43.5	102.4
	129.9	12.0	37.1	76.5
	60.5	8.3	26.5	71.3

b) Double fitting of T_1-Torchia values (s)

PET sample				
ppm		as received	140 °C	200 °C
	162.5	23.9	79.3	437.7
		-	0.6	18.3
	132.7	24.2	70.4	227.0
		-	1.1	17.5
	129.9	19.6	54.8	143.4
		-	1.6	11.6
	60.5	12.7	48.3	119.6
		-	1.6	5.3

As evident from the table, T_1 shows extremely different values from the as received sample with respect to the sample treated for 60 minutes at 140°C. Resolving the decay curves of this last sample with a binomial function a significant improvement in the standard deviation is evidenced, as well as the appearance of two values for T_1

explainable respectively by the existence of a rigid and of an amorphous structure.

Clearer information in order to investigate PET materials can be obtained also through spin diffusion measurements obtained trough spin-exchange term in the Hamilton Integral describing the body proton-proton dipolar coupling. In the PET case spin diffusion measurements are easier because of the fact that they are of at least two orders of magnitude bigger of those performed in order to reach the equilibrium and therefor magnetisation remains constant together with the fact that the diffusion coefficients in the different regions of the material are extremely similar.Table 3 shows the $T_{1\rho}$ values for the samples heated for 60 min.

Table 3 - $T_1\rho$ of the PET samples (ms)

Air annealed PET					
ppm	as received	140 °C	160 °C	180 °C	200 °C
162.5	5.0	14.3	14.6	17.8	19.5
132.7	5.1	12.6	12.3	15.5	16.7
129.9	5.7	14.6	14.7	17.8	19.1
60.5	5.9	16.3	16.0	19.8	22.3
Microwaves annealed PET					
ppm		140°C	160 °C	180 °C	200°C
162.5		6.2	14.1	13.9	14.4
132.7		9.7	12.5	12.5	12.9
129.9		10.9	14.2	14.5	14.4
60.5		12.5	15.9	16.0	16.0

From the spin diffusion values the increment of the intensity of the signal at 60.5 ppm with increasing the annealing temperature in the conventional treatment a clear increment of the dimension of the ordered phase is pointed out.

Comparing the $T_{1\rho}$ values for the single signal of the spectrum of the as received product the expected equivalence of the values is observed, typical of homogeneous systems, suggesting the presence of a phase of little dimensions. In the remaining annealed samples on the contrary, variations – particularly evident in the signal of the aromatic groups at 132.7 ppm - suggest significant dishomogeneities in the structure of materials.

Concluding remarks

In order to achieve innovation as a strategic element of product diversification and of processes competition, the present research allowed us to verify that PET textiles can be thermosetted by microwaves.

The effectiveness of the treatment depends on the operative conditions chosen, particularly on temperature and annealing time.

Microwave thermosetting seems to be less effective than the conventional one, but the use of single mode furnaces should ensure a better power density and therefore increase the effectivness of the treatment.

[1] D.A Torchia, *J.Magn. Reson.* **1978**, 30, 613.
[2] A Abragam, *The Principles Of Nuclear Magnetism*; Oxford University Press, Oxford, 1961.
[3] R. Voekel,, *Angew. Chem.* **1988**, 100, 1525.
[4] G. Groenickx, H. Berghmans and G.J Smets, *J.Polym. Sci., Polym. Phys. Edn.* **1976**, 14, 591.
[5] G.C Alfonso, E. Pedemonte and L. Ponzetti, *Polymer* **1979,** 20, 104.
[6] A Jeziorny, *Acta Polymerica,* **1986**, 37, 237.
[7] K. Scmidt-Rohr, W. Speiss, *Multidimensional Solid State NMR and Polymers*, Academic Press, London, 1994.
[8] J.R. Havens, D.L Vanderhart, *Macromolecules* **1985,** 18, 1663.
[9] K Schmidt, W.H. Zumbulyadis *Science,* **1998**, 280, 714.

[10] J.M Huang,., Chu, P.P. and F.C. Chang, *Polymer,* **2000**, 41, 1741.
[11] M.D. Sefcik, J. Schaefer, E.O. Stejaskal and R.A. Mckay **1980**, *Macromolecules* 13, 1182.

The Lamellar Distribution in Isotactic Polypropylene Modified by Nucleation and Processing

Anna Romankiewicz, Tomasz Sterzynski

Poznan University of Technology, Polymer Group, PL 60-965 Poznan, Poland

Summary: An isotactic polypropylene was modified with specific α - phase and β - phase nucleation agents. The modification by processing during extrusion was realised by means of molten state elongation with various stretching ratios λ between 7.11 and 1.67. The lamellar thickness distribution was evaluated for non - elongated and elongated samples using the DSC melting curves, by an application of the general Thomson - Gibbs equation. A significant influence of heterogeneous nucleation on the lamellar distribution was confirmed. The morphological effects are visible as a shift of the maxima of the lamellar distribution and as the elongation and nucleation dependent appearance of lamellae with smaller (below 12 nm) and bigger (above 25 nm) thickness.

Introduction

The lamellae are the structural elements produced by growth of a three dimensional order during cooling of melt of semi crystalline polymers. The fold - chain fibrils forming the lamellae are produced by a radial growth of crystal systems, with less or more small - angle branching of fibrils [1-3]. Bigger units like spherulites may be found in most of semi crystalline polymers, especially, in polymers characterized by lower crystal growth rate such as polypropylene (ca. 20 μm/min), or polyethylene oxide (200 μm/min) but also in polyethylene (growth rate >1000 μm/min) [4,5]. Usually the thickness of the lamellae is of the order of about 10 nm, but by an isothermal crystallisation an overlaid of single lamellae may create the lamellae packs with a thickness of about 100 nm. If the thickness exceeds 1 μm the lamellae may be observed directly by means of the optical microscopy [6].

Depending on the polymer chemical structure and the macromolecular mobility, within the structure unit, two direction of fold - chain fibril growth: radial [R] and tangential [T] may usually be observed. For all semi crystalline polymers the lamellae thickness presents a considerable dependence on the crystallisation conditions, especially on the crystallisation temperature and the annealing time. For a sufficient high temperature and

 CCC 1022-1360/00/$ 17.50+.50/0

time the lamellar thickness may increase even ten times [7-9] comparing with low temperature and non-isothermal crystallization conditions.

The nature of the lamellar morphology may usually be directly observed by polarised light optical microscopy (MPO) or by scanning electron microscopy (SEM). The average lamellar thickness may be detected by means of small angle X - ray scattering (SAXS) [6] or by the differential scanning calorimetry (DSC) [10-14]. In the case of SAXS measurements, the position of the maximum on the diffraction curve $Is^2 = f(s)$, where I is the diffraction intensity and s - the scattering vector, allows to evaluate the average value of the long period, i.e. the distance between two crystalline layers in a multi-layer lamella model. By the DSC measurements the lamellar thickness l may be estimated using of the Thomson - Gibbs relationship.

Lamellar Thickness by DSC

For this procedure a hypothesis relating the lamellae distribution to the form the DSC curve $\Delta H = f(T)$ is taken into account [10-13]. It is known that the temperature affects the crystallisation kinetics, and from other side the lamella thickness affects the melting point of the polymer crystals. This relationship was expressed in a form of the Thomson-Gibbs equation [14]:

$$T_m = T_m^0 \left(1 - \frac{2\sigma_e}{l \Delta H_f} \right) \qquad (1)$$

where:

T_m - melting temperature of the investigated polymer,

T_m^0 - the equilibrium melting temperature,

l - the lamella thickness (longitudinal dimensions of the crystal),

ΔH_f - the melting enthalpy of a perfect crystal,

σ_e - free surface energy of the end faces at which chains fold.

The thickness distribution of the crystalline lamellae, related to the DSC thermograms, may be expressed by the following equation:

$$f(l) = \frac{1}{M} \frac{dM}{dl} \qquad (2)$$

where the lamella thickness of a polymer is related to the melting temperature T_m of a single crystal by the equation:

$$l = \frac{2\sigma_e T_m^0}{\Delta H_f \left(T_m^0 - T_m\right)} \qquad (3)$$

and

$$dM = \frac{dE}{dT}\frac{dT}{\Delta H_f}\rho_c \qquad (4)$$

where:

ρ_c - the density of the crystal phase,

dE - the energy necessary to melt the crystalline mass dM in the temperature range between T and $T+dT$.

From eq. (3) and (4) follows:

$$\frac{dM}{M} = \frac{dE}{dT}\frac{dT}{\Delta H_f M}\rho_c \qquad (5)$$

where the temperature increase:

$$dT = dl\frac{\left(T_m^0 - T_m\right)^2}{T_m^0}\frac{\Delta H_f}{2\sigma_e} \qquad (6)$$

and the thickness distribution may be expressed as follows:

$$\frac{1}{M}\frac{dM}{dl} = \frac{dE/dT\left(T_m^0 - T_m\right)^2 \rho_c}{2\sigma_e T_m^0 M} \qquad (7)$$

The evaluation of the following equation is based on the DSC measurements:

$$\frac{dE}{dT}\frac{1}{M}\left[\frac{J}{{}^oC}\frac{1}{kg}\right] = Heat.Flow\left[\frac{J}{kg \cdot \min}\right] / V_{cooling.or.heating}\left[\frac{{}^oC}{\min}\right] \qquad (8)$$

The *Heat Flow* is the value of the ordinate on the original DSC curve with subtracted base line. The parameters for the description of the crystallization of the monoclinic (α-form) and hexagonal (β-form) of an isotactic polypropylene, according to [15-17], are presented in Table 1.

Table 1. The parameters characterizing the crystallization of α-iPP and β-iPP.

	T_m^0 [K]	ΔH_f [J/cm^3]	σ_e [J/cm^2]	ρ_c [g/cm^3]
α-phase iPP	464.0[14]	196[14]	102.9[14]	0.936[15]
β-phase iPP	449.2[16]	177[16]	55.2[16]	0.921[15]

Experimental Part

Material

A commercial iPP NOVOLEN 1100H, produced by BASF (Germany), was used in this study. The density is 910 kg/m^3 and the melt flow rate (MFI 230°C, 2.16 kg) = 2.4 cm^3/ 10 min.

Nucleating agents: 1, 3: 2, 4 – bis (3, 4-dimetylobenzylideno) sorbitol (DMDBS) (Milliken, Belgium), with melting point T_m =277°C and crystallization temperature T_{cr} = 214°C, was applied as a specific α nucleating agent, and N, N' – dicykloheksylo – 2, 6 – naftaleno dikarboksy amide (NJ), (Palmarole Ltd, UK) with melting point T_m = 387°C and crystallization temperature T_{cr} = 137°C, as a β - form nucleate.

Sample preparation

The nucleates were mixed with the pallets via powder sprinkling on the surface of the granules. The homogenization was performed using a single screw extruder FAIREX (ϕ = 30 mm and L/D = 24) operating at the barrel temperatures of 25°C, 140°C, 180°C, 185°C, and the die at 195°C and 190°C, respectively. The screw rotation speed was 30 rpm. A powder mixing and molten state extrusion was used to prepare two master batches containing 0.5 wt. % of both, the α- phase (DMDBS) and β- phase (NJ) nucleating agents. This procedure was succeed by solution of the master batch with pure iPP, with the aim to achieve the concentration of nucleating agents (NA's) in the range from 0.001 wt. % to 0.5 wt. %.

For a better distribution of the NA's in the iPP matrix, a double melting mixing procedure was applied for each composition.

Molten state elongation

The molten state elongation was completed between the extrusion die and the entrance to the cooling device, by means of a profile extrusion with a flat die 2 x 20 mm, and a pulling device operating with various drawing rates. By variation of the pulling rate, stretching ratios (λ) of the profile between λ = 1.67 and λ = 7.11 was achieved.

Differential scanning calorimetry (DSC)

For the thermal characterization of the polymeric materials the DSC Netzsch 200 thermal analysis system was used. The average weight of the samples was between 5 and 8 mg. Following DSC temperature program was applied: first heating, in the range from 30°C to 200°C, with a rate of 10°C min^{-1}, maintenance at this temperature for 5 min, cooling to 40°C with a cooling rate of 5°C min^{-1}, maintenance at 40°C during 5 minutes and second heating up to 200°C with a rate of 10°C min^{-1}.

For all specific nucleated and elongated iPP samples following parameters were determined: the melting temperature T_m, the crystallization temperature T_{cr} as well the temperature dependent heat of fusion ΔH_f (shape of the DSC first melting curve). As the aim of these studies was to relate the iPP lamella structure to the nucleation and processing history, the evaluation of the lamellar distribution was performed always for the first DSC melting curve. For comparison, also samples without nucleation, neither elongation were examined.

It is known that the first heating scan of a DSC is affected by many errors, which can significantly influence the reported results. The most important is the heat exchange coefficients between sample and holder, as well between holder and oven [18]. In order to minimize this error and to assure an adequate heat flow, each of samples were flat adhered to the DSC pans.

To examine the structure of polypropylene modified by nucleation and processing, the DSC samples were taken through the whole thickness of the film, in the middle of flow. On this way the results may be comprehend as typical mean values of processing and nucleation induced iPP structure modification. In order to minimize the measurement error and to determinate the average value of T_m and ΔH_f the DSC melting procedure, performed always in the same conditions, was repeated for a number of samples. We have observed an adequate repetition of the DSC results, e.g. of the melting temperature and of the corresponding enthalpy.

Results

DSC melting curves

The DSC melting process by an isotactic polypropylene may be strongly influenced as well by the thermal and mechanical history [19] as by the composition of the polymeric system. Even a very low content of specific nucleating agents may significantly modify the DSC melting curve, as it was already shown in the case of quinacridone and other additives [4,5,20,21]. Same results were attained in the reported case where the β - phase in iPP was produced by the addition of NJ [22,23].

The following curves reveal to iPP modified by heterogeneous nucleation with various content of the specific nucleating agents, produced by an extrusion homogenization and following palletizing. The DSC melting curves for the iPP nucleated with DMDBS are presented on Figure 1.

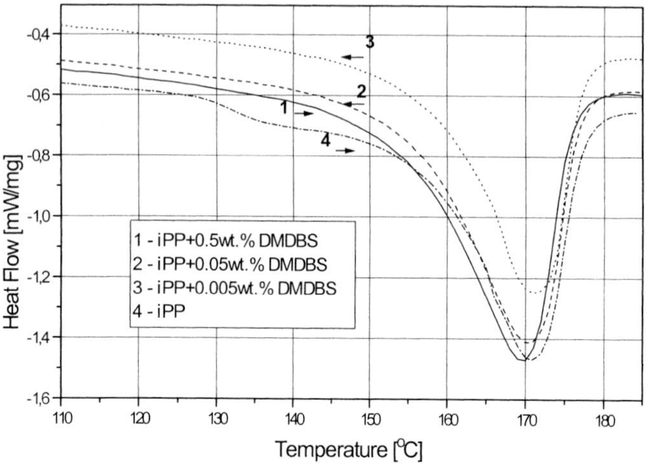

Figure 1. The melting curves of iPP nucleated with DMDBS.

The melting curves for the iPP nucleated with DMDBS present almost the same shape for all compositions, independent on the quantity of the nucleating agent added. On Figures 2 and 3 the DSC melting curves for the iPP nucleated with various content of NJ are shown where the curves on Figure 2 correspond to the first melting and on Figure 3 to the second melting.

The comparison of values of T_m and ΔH_f for iPP without and with different content of nucleating agents is presented in Table 2. A small shift of melting temperature observed by DSC first melting run may be related to somehow modified thermal stability of various samples or/and to certain measurement error [18], a problem which was already discussed before.

The difference in the run of the DSC curves in both cases may be related to various cooling conditions. The DSC melting curves presented on Figure 2 were recorded for rapidly cooled samples, produced by molten state extrusion followed by palletizing and cooling in a water bath. On the contrary, curves on Figure 3 represent the second DSC melting, e.g. after cooling of the samples in the DSC apparatus with a controlled cooling rate of 10 °C min^{-1}. It's evident that during fast cooling (case on Figure 2), even if the appropriate additives are used, the development of the α - phase is predominating [4,24], and therefore a very small or even any maxima typical for the β - phase may be observed. By much lower cooling rates (case on Figure 3) a significant development of the β - phase may be noted. In this case two separate maxima corresponding to the melting of the β - phase (by about 150°C) followed by melting pick of the α -phase (167°C) usually may be observed. The explanation of the existence of both picks and the melting and recrystallization phenomena was proposed and discussed by Varga [4,5,20]. The ratio between the heights of both picks (β and α - phase) corresponds to the β - phase content, effect that will be presented elsewhere [23].

Table 2. Melting point and melting enthalpy of iPP with and without nucleating agents (first heating).

	T_m [°C]	ΔH [J/g]
iPP	170.8	95.4
iPP+0.005 wt. % DMDBS	170.7	83.2
iPP+0.05 wt. % DMDBS	170.4	81.2
iPP+0.5 wt. % DMDBS	170.0	92.0
iPP+0.005 wt. % NJ	171.1	87.1
iPP+0.05 wt. % NJ	169.9	83.5
iPP+0.5 wt. % NJ	169.8	77.6

The example of WAXS patterns for non-nucleated ("pure") and specific nucleated iPP are shown on Figure 4. For "pure" iPP sample a WAXS spectrum corresponding to α – phase polypropylene may be seen. On the contrary, the addition of the NJ – nucleating agent leads to the creation of the β – phase iPP with characteristic maxima corresponding to the pseudo-hexagonal crystal structure. In the reported case a relative content of the β-form reached values of about 99%, for the NJ concentration of 0.5 wt.%.

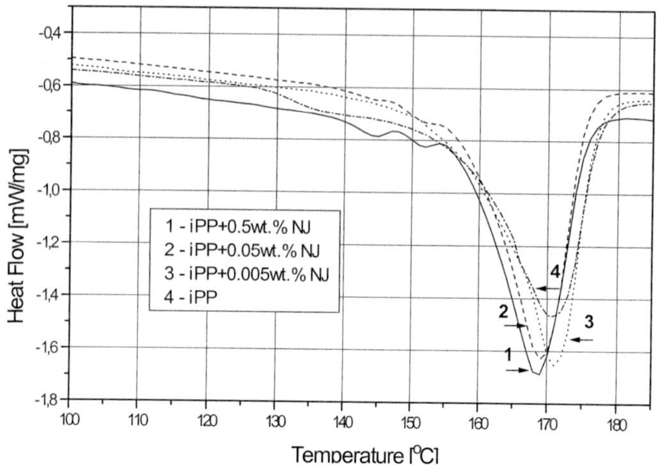

Figure 2. The melting curves of iPP with NJ (first melting).

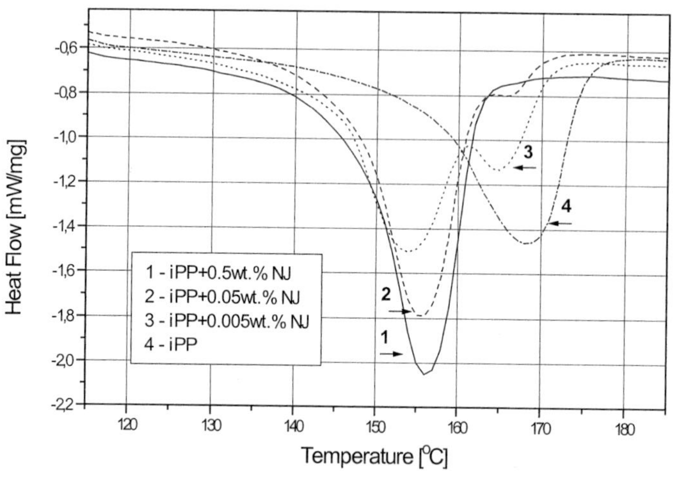

Figure 3. The melting curves of iPP nucleated with NJ (second melting).

The following results reveal to samples produced by melt stretching in an extrusion process. The DSC melting curves for non-nucleated and nucleation modified iPP samples, processed with various molten state stretching ratios, are shown on Figures 5 to 7. As it may be seen a higher elongation ratio leads to a duplication of the maximum, an effect that may indicate an elongation dependent creation of a bi - modal lamellar thickness distribution.

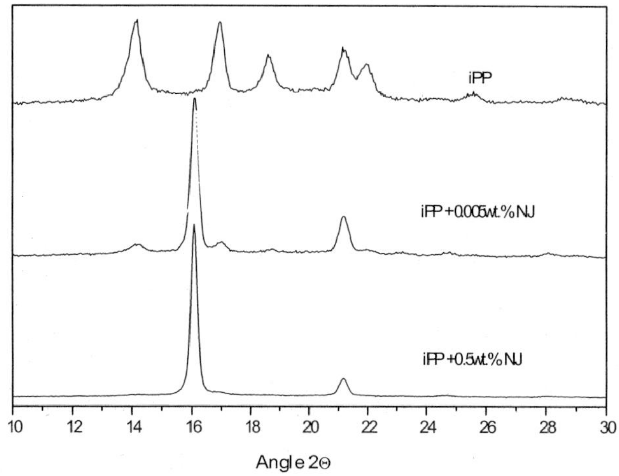

Figure 4. The WAXS patterns of iPP "pure" and with NJ.

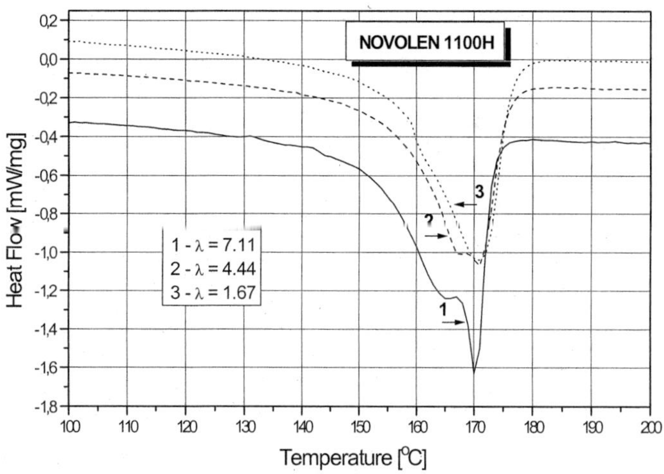

Figure 5. The melting curves of non-nucleated iPP samples processed with different λ.

The stretching ratio applied in this study led to different orientation degree in extruded products, as determined by FTIR spectroscopy [23].

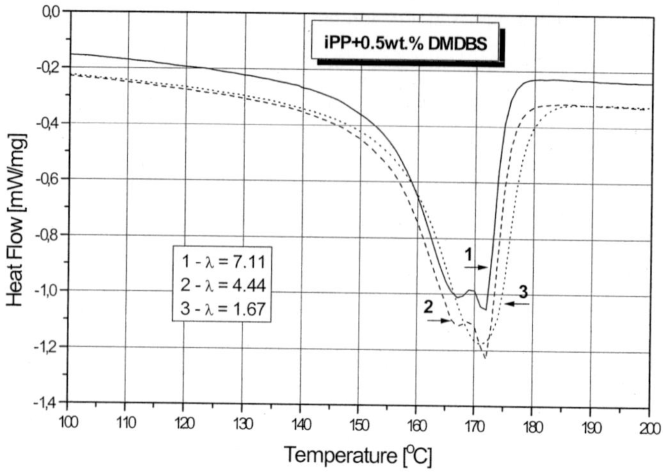

Figure 6. The DSC melting curves of iPP samples nucleated with DMDBS and processed with different stretching ratios.

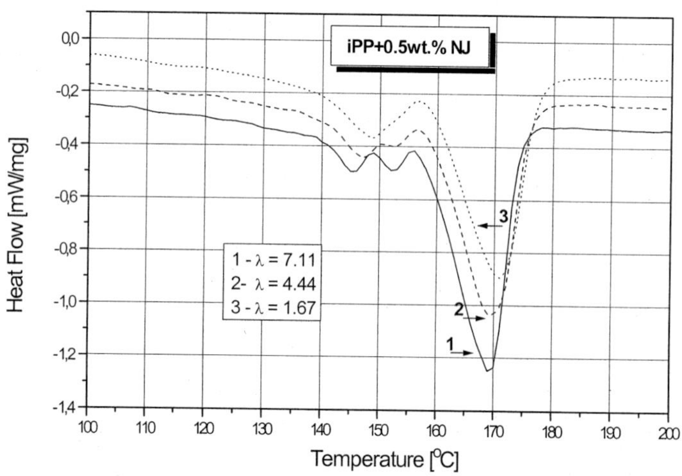

Figure 7. The melting DSC curves of iPP nucleated with NJ and processed with different stretching ratios.

Nucleation dependent lamellar distribution

The lamellar distribution function $f(l) = \frac{1}{M}\frac{dM}{dl}$, for various compositions of iPP samples and elongation ratios, are presented on Figures 8 to 12. For the estimation of the nucleation dependent lamellar distribution only the equilibrium temperature for the monoclinic iPP was taken into account, as the performance of the DSC peak separation for the α/β phase iPP isn't convinced. Therefore, the comparison of the lamellar distribution, for samples produced the by varied nucleating agent content and by various melt stretching, was achievable. On Figure 8 the lamellar distribution for pure iPP and iPP with 0.5 wt. %, 0.05 wt. % and 0.005 wt. % of DMDBS (α - phase nucleating agent) is presented.

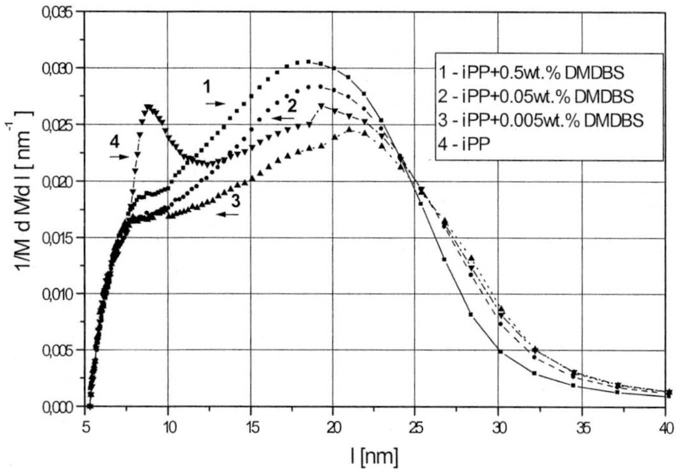

Figure 8. The lamellar distribution for pure iPP and iPP with 0.5 wt. %, 0.05 wt. % and 0.005 wt. % of the DMDBS.

The changes of the concentration of the nucleating agent lead to certain modification of the lamellar distribution. In this case two effects may be observed and discussed separately, a shift of the maximum corresponding to the most probable lamella value, and the existence of bigger lamellae with an average value above 25 nm. As it follows from figure 8, an increase of the DMDBS content leads to somehow lower average lamellae values, (maximum at about 22 nm for 0.005 wt. % and at about 18 nm for 0.5 wt. %) and more homogeneous lamellar distribution, e.g. a much smaller participation of bigger lamellae, over 25 nm. This kind of concentration dependent lamellae distribution may be compared to the relationship between the modification of the crystallization temperature T_{cr} and the DMDBS content, where a similar influence of the DMDBS content on the T_{cr} was found [23].

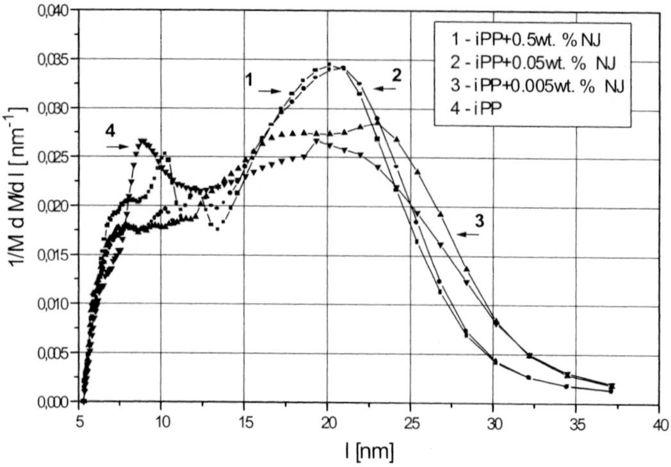

Figure 9. The lamellar distribution for pure iPP and iPP with 0.5 wt. %, 0.05 wt. % and 0.005 wt. % of NJ.

For the iPP with the β - phase nucleating agent - NJ (Figure 9) similar observation, as for iPP modified with DMDBS, may be done (for comparison the lamellar distribution was evaluated based on the first melting DSC curve). A concentration dependent shift of the maximum towards lower lamellae values and a clear change in the form of the curve, with a significant part of higher lamellae values (about 25 nm), for the lowest NJ-content, was noted. This kind of relationship may also be attributed to the NJ - concentration dependent changes of the crystallization temperature - T_{cr}, where even a low NJ-content leads to a significant increase of T_{cr} - an effect which was already observed and confirmed in our research [22,23]. Additionally, a certain amount of smaller lamellae, with average values between 7 and 15 nm, may be attributed to the considerable β - phase existence in these iPP samples (the content of the hexagonal phase determined by WAXS was between 89 and 99% [23]).

For comparison with presented results also an isothermal crystallization of "pure" iPP samples by 122°C, 124°C, and 127°C was realized by the DSC procedure. A satisfactory agreement of the lamellar distribution for an isothermal crystallization with curve 4 on Figure 9 was found.

Molten state elongation induced modification of the lamellar distribution

On following figures the influence of molten state stretching on the lamellar distribution, in iPP samples without and with nucleating agents, is presented. The curves on Figure10 represent the lamellar distribution of "pure" iPP elongated with various stretching ratios λ.

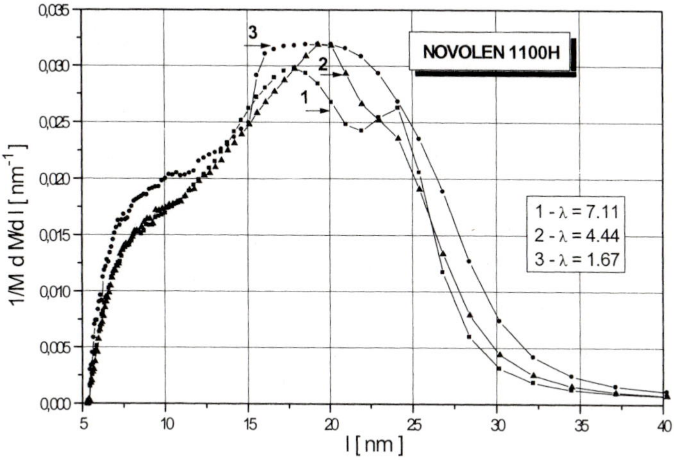

Figure 10. The lamellar distribution of pure iPP elongated with various stretching ratios λ.

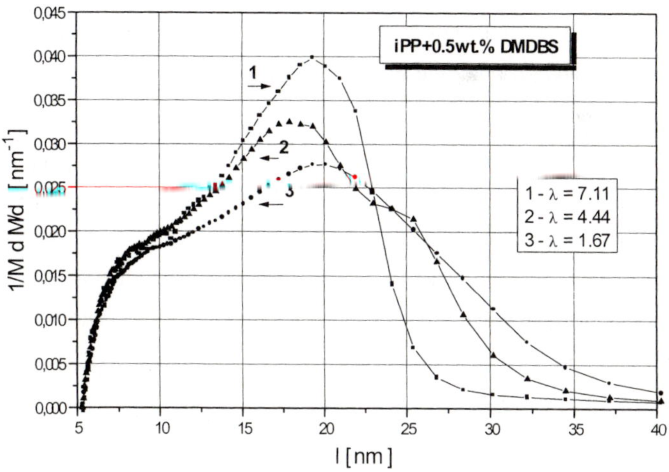

Figure 11. The lamellar distribution for iPP with 0.5 wt. % of DMDBS elongated with various stretching ratios λ.

For the highest λ - value by uni-axial elongation of "pure" iPP a relative homogeneous lamellar distribution, with a bimodal distribution may be observed. On the contrary, a rather uniform curve in the middle part, with somehow bigger part of broaden lamellae, over 25 nm, was noted if a lower stretching ratio (λ = 1.67) was applied.

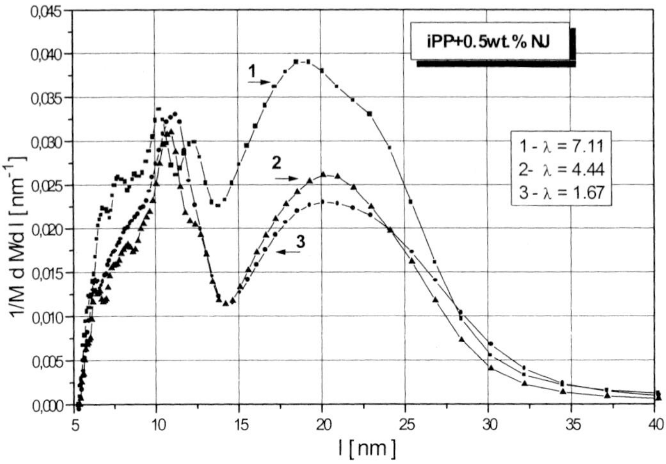

Figure 12. The lamellar distribution for iPP with β -phase nucleating agent, stretching ratios λ = 1.67, 4.44 and 7.11.

With increasing stretching ratio λ. a more narrow run of the lamellar distribution function may be observed for the specific α - nucleated iPP samples (Figure 11). For the stretching ratio λ = 7.11 the distribution function $f(l) = \frac{1}{M}\frac{dM}{dl}$ presents a smooth curve with a clear maximum for l = 20 nm. On the contrary, for lower uni – axial deformation the lamellar distribution is much broader, with a bimodal distribution for λ = 4.44. The lamellae with a mean thickness above 25 nm are observed only for lower stretching ratios. Figure 12 presents the distribution function (equation 2) for the iPP samples nucleated with β -phase nucleating agent and elongated to λ = 1.67, 4.44 and 7.11. Similar, like in Figure 11, a significant maximum for the highest stretching ratio was observed, where for lower λ a fewer developed maximum of the lamellar distribution function for β – nucleated iPP may be noted. For this type of iPP modification also a significant participation of lamellae with mean thickness values in the range between l = 7 nm and l = 12 nm, related to the existence of β - phase lamellae may be observed. The probably reasons for the change of the average lamella value are:

- dependent of the crystallization ratios on the specific nucleation of iPP, or/and
- temperature dependent increase of crystal growth of monoclinic and hexagonal phase of iPP in defined temperature range [5].

Conclusions

A significant influence of the heterogeneous nucleation on the lamellar distribution in the iPP samples, deformed uni-axial in a molten state, was confirmed. The changes of the lamellar distribution function may be attributed to the nucleation-induced modification of the crystallization kinetics. In relationship with molten state processing, the use of specific additives (nucleating agents) plays a dominating role as well by the changes of the germination density, as by the creation of various lamellae types and by modification of its radial growth ratio.

Acknowledgement

Project partly supported by Polish Research Committee KBN 7.T08E.047.17

References

[1] B. Wunderlich, *Macromolecular Physics,* Academic Press, New York (1973).
[2] D. Trifonova, J. Varga and G. J. Vancso, *Polymer Bulletin*, 41, (1998), p. 341.
[3] H.-G. Elias, *Macromolecules 1, Structure and Physics,* Plenum Press, New York (1984).
[4] J. Varga, *J. Thermal Analysis*, 35, (1989), p. 1891.
[5] J. Varga, *J. Mater. Sci.*, 27, (1992), p. 2557.
[6] T. Sterzynski, "*Lamella dimension and distribution*", [in:] *Polypropylene: an A-Z*, Ed. J. Karger-Kocsis, Kluwer Academic Publisher, Dordrecht (1999), p. 374.
[7] K. Yamada, S. Matsumoto, K. Tagashira and M. Hikosaka, *Polymer*, 39, (1998), p. 5327.
[8] M. Aboulfaraj, B. Ulrich, A. Dahoun and C. G'sell, *Polymer,* 34, (1993), p. 4817.
[9] T. Sterzynski, P. Calo, M. Lambla and M. Thomas, *Polym. Eng. Sci.*, 37, (1997), p. 1.
[10] Y. L. Huang and N. Brown, *J. Polym. Sci.*, 29, (1991), p.129.
[11] M. Alonso, J. C. Rodriguez – Cabello, J. C. Merino and J. M. Pastor, *Macromol. Chem. Phys.*, 197, (1996), p. 3269.
[12] C. J. G. Plummer and H.-H. Kausch, *Polymer Bulletin,* 36, (1996), p. 355.
[13] J. C. Rodriguez – Cabello, M. Alonso, J. C. Merino and J. M. Pastor, *J. Appl. Polymer Sci.*, 60, (1996), p. 1709.
[14] V. A. Bershtein and V. M. Egorov, *Differential Scanning Calorimetry of Polymers. Physics, Chemistry, Analysis,* Ellis Horwood, London (1994).
[15] Z. Bartczak and A. Galeski, *Polymer,* 31, (1990), p. 2027.
[16] J. Karger – Kocsis, J. Varga and G. W. Eherenstein, *J. Appl. Polymer Sci.*, 64, (1997), p. 2057.
[17] G. Shi, X. Zhang and Z. Qiu, *Makromol. Chem.*, 193, (1992), p. 583.
[18] G.Eder, H.Janeschitz-Kriegl, The Polymer Processing Society, Ninth Annual Meeting, Manchester (1993) p. 452
[19] T. Sterzynski, H.Oyased, to be published

[20] J. Varga, *J. Thermal Analysis*, 31, (1986), p. 165.
[21] T. Sterzynski, M. Lambla, F. Georgi and M. Thomas, *Intern. Polymer Processing XII*, 1, (1997), p. 64.
[22] A. Romankiewicz and T. Sterzynski, *World Polymer Congress IUPAC* MACRO 2000, Warsaw (2000) p. 966.
[23] A. Romankiewicz, Phd Thesis, Poznan University of Technology, Poznan 2001.
[24] B. Fillon, A Thierry, J. C. Wittmann and B.Lotz, *J. Polymer Sci., Part B, Polymer Phys.*, 31, (1993), p. 1407.

Macromol. Symp. ***180,*** *257–266 (2002)*

Thermal Degradation of Two Classes of Block Copolymers Based on Poly(lactic-glycolic acid) and Poly(ε-caprolactone) or Poly(ethylene glycol)

Fabio Bignotti°, Maurizio Penco°, Luciana Sartore°, Salvatore D'Antone^, Alberto D'Amore°, Gloria Spagnoli°*

°Dipartimento di Chimica e Fisica per l'Ingegneria e per i Materiali, Via Valotti 9, 25133 Brescia, Italy

^Dipartimento di Chimica e Chimica Industriale, Via Risorgimento 35, 56100 Pisa, Italy

Summary: Thermodegradative investigations of two classes of multi-block copolymers containing poly(D,L-lactic-glycolic acid) (PLGA) and either poly(ethylene glycol) (PEG) or poly(ε-caprolactone) diol-terminated (PCDT) segments were performed. In particular, the influence of the type and length of the segments as well as of the molar ratio between the D,L-lactic acid (LA) and glycolic acid (GA) residues was investigated at 180°C in air by viscometry, FT-IR analysis and isothermal thermogravimetry. The thermal oxidative degradation of these materials is largely affected by the LA/GA ratio, a higher LA content generally imparting higher stability. The FT-IR analysis suggests that, depending on the composition of the PLGA segments, degradative processes are triggered which can lead to a preferential degradation of the blocks.

Introduction

Poly(lactic-glycolic acid) (PLGA) belongs to a family of biocompatible and bioerodible polymers currently employed in the biomedical field as surgical materials, matrices for bone reparation, scaffolds for tissue regeneration, drug delivery systems [1-5] *etc.* Polymers with variable lactic acid (LA) over glycolic acid (GA) molar ratios, or containing D,L-lactic acid mixtures, can be prepared to modulate the properties of these materials and to optimise their performance in view of the specific involved applications. To that purpose, PLGA have also been copolymerised with other polymers [4,6]. Such materials are often subjected to thermal treatments, for instance during moulding or sterilisation, which can induce polymer degradation, thus affecting both their mechanical properties and their performance in vivo [7-11]. Therefore, the knowledge of their calorimetric properties and thermal stability is of great

 CCC 1022-1360/00/$ 17.50+.50/0

interest from the technological point of view.

In recent years we have developed a procedure to synthesise poly(ester-carbonate)s with variable structures. In particular, our attention has focused on two classes of multi-block copolymers containing segments of poly(D,L-lactic-glycolic acid) and of diol-terminated poly(ε-caprolactone) (PCDT) or poly(ethylene glycol) [12-14]. The copolymers thus obtained (PLGA-PCDT and PLGA-PEG, respectively) contain blocks linked to each other through carbonate or ester linkages:

$$\left[\left(\underset{|}{\overset{R}{}}\!\!CH{-}COO \right)_n R'{-}O{-}COO \right]_x$$

where:

$R = H$ or CH_3

$R' = \left[(CH_2)_5{-}COO \right]_a CH_2CH_2 \left[OCO{-}(CH_2)_5 \right]_b$ in **PLGA-PCDT**

$R' = -CH_2 {-}\left(CH_2OCH_2 \right)_m{-} CH_2-$ in **PLGA-PEG**

Previous investigations showed that semicrystalline or amorphous materials endowed with different calorimetric properties and susceptibility to hydrolytic degradation can be obtained varying the copolymer composition [15-16]. More recently, their thermal stability was also studied by TGA, from room temperature to 650°C, both under nitrogen and air flow [17]. It was concluded that PLGA-PCDT and PLGA-PEG do not exhibit any noticeable weight loss at temperatures lower than about 200–250°C. However, processes that lead to polymer degradation might be active at those temperatures even if they could not be detected by TGA for the absence of appreciable volatilisation under the analytical conditions adopted (scanning rate = 20°C/min, flow rate = 20 ml/min). Therefore, the aim of this paper is to report the results of more extensive studies on the thermal stability of the above copolymers. To that purpose they were annealed in air at 180°C and their behaviour was monitored by isothermal TGA, viscometry and FT-IR analysis. For comparison, the homopolymers constituting the blocks were also investigated by the same techniques.

Experimental Part

Instruments

TGA analysis was run by a Mettler TC11 Thermal Analyser Processor equipped with a Mettler TG50 microbalance, heating samples of about 20 mg into alumina crucibles at 180 °C under air flow (10 ml/min). The FT-IR spectra were recorded on a Jasco 5300 FT-IR spectrophotometer by casting from chloroform on KBr windows. Viscometric determinations were performed in chloroform at 30°C by means of Ubbelohde viscometers.

Materials

PCDT oligomers, PEO and high molecular weight PLGA50/50 ($\overline{M}_w$ = 50,000-75,000), PLGA75/25 ($\overline{M}_w$ = 75,000-120,000) and PCL ($\overline{M}_w$ = 65,000, $\overline{M}_n$ = 42,500) were purchased from Aldrich Co. PLGA oligomers, PLGA-PCDT, PLGA-PEG and PLA-PEG block-copolymers were synthesised as previously described [12-14].

Film preparation

Materials used for viscometric and FT-IR studies were films obtained by casting from chloroform solutions (c = 70 mg/ml). Each film was prepared pouring exactly 1 ml of solution into a 15 x 100 mm test tube and allowing the solvent to evaporate first at room pressure and then at 0.1 mmHg up to constant weight. For each material a number of replicate films were prepared. Samples of the desired material were annealed in an oven at 180°C. At intervals (after 15, 30, 60, 120 or 240 minutes), they were withdrawn from the oven, put in a desiccator, allowed to cool to room temperature and analysed.

Results and Discussion

Copolymers with different LA/GA ratios and different segment lengths were investigated to analyse the influence of structure on the degradative behaviour of the materials. Segments of poly(D,L-lactic acid) (PLA) and of PLGA with two LA/GA molar ratios typically employed in the biomedical field (50/50 and 75/25) were considered. For every ratio, PCDT segments with molecular weight 530, 1250 and 2000, respectively, or PEG segments having molecular weight 2000 were used. The composition of the copolymers investigated and the molecular weight of segments are summarised in Table 1. The molecular weight values were obtained by size exclusion chromatography (SEC) using a polystyrene calibration curve.

Table 1. Molecular weight characterisation of block copolymers.

Code	Starting oligomers				Block copolymers			
	PLGA		Diol		$[\eta]^{a)}$	$\overline{M}_n$ [b)]	$\overline{M}_w$ [c)]	D[d)]
	M_r[e)]	$\overline{M}_n$ [f)]	Type	$\overline{M}_n$	(dl/g)	$x10^{-3}$	$x10^{-3}$	
PLGA50/50 - PCDT530	1:1	1920	PCDT	530	0.40	8.4	31.5	3.7
PLGA50/50 - PCDT1250	1:1	1920	PCDT	1250	1.20	22.0	114.7	5.2
PLGA50/50 - PCDT2000	1:1	1920	PCDT	2000	0.82	27.0	113.7	4.2
PLGA75/25 - PCDT530	3:1	2830	PCDT	530	0.81	18.0	97.1	5.4
PLGA75/25 – PCDT1250	3:1	2830	PCDT	1250	1.49	26.6	133.7	5.0
PLGA75/25 – PCDT2000	3:1	2830	PCDT	2000	1.45	22.8	123.9	5.4
PLGA50/50-PEG2000	1:1	1920	PEG	2000	1.55	197	370	1.9
PLA-PEG2000	-	3050	PEG	2000	1.63	119.5	301.2	2.5

a) Intrinsic viscosity in chloroform at 32 °C.

b) Number-average molecular weight evaluated by SEC using polystyrene narrow standards.

c) Weight-average molecular weight evaluated by SEC using polystyrene narrow standards.

d) Polydispersity index $\overline{M}_w/\overline{M}_n$.

e) LA:GA molar ratio.

f) Determined by end-group titration.

Isothermal TGA

The copolymers containing PCDT or PEG segments of molecular weight 2000 were analysed by isothermal TGA in air, at 180°C, for 240 min. The traces of PLGA-PCDT2000 copolymers are reported in Figure 1 along with those of the corresponding homopolymers. PLGA75/25 and PLGA50/50 exhibit comparable weight loss (WL) while PCL is more stable. Interestingly, the volatilisation rate of PLGA75/25-PCDT2000 is very similar to that of PCL, whereas it is much higher in PLGA50/50-PCDT2000. Figure 2 shows the results obtained for the copolymers with PEG and parent homopolymers. PEO undergoes rapid degradation when heated in air, in agreement with data reported in literature [18]. PLGA-PEG copolymers show a higher WL than PLGA-PCDT, but it is less remarkable than in PEO. In addition, while the WL of PLA-PEG2000 increases almost linearly with time, a more complex trend is exhibited by PLGA50/50-PEG2000, suggesting that different degradative processes are active in these materials. Again, the copolymer with a lower LA content degrades faster, even if the difference becomes less pronounced with time. Therefore in both families of copolymers the thermoxidative stability seems to be enhanced by the presence of a higher LA content.

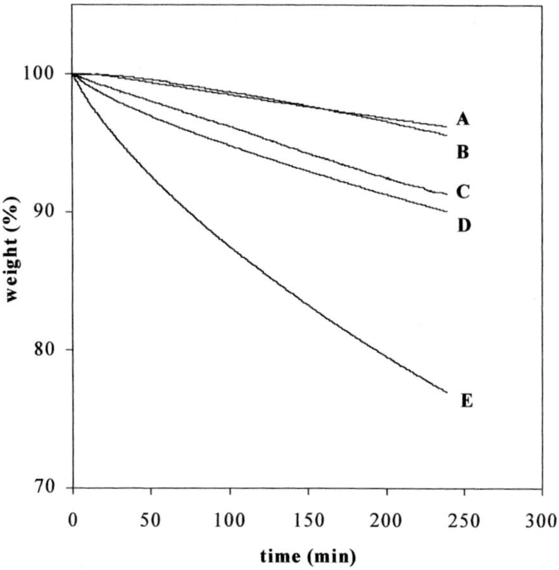

Figure 1. Weight percentage *vs.* heating time at 180°C in air for PLGA75/25-PCDT2000 (A), PCL (B), PLGA50/50 (C), PLGA75/25 (D) and PLGA50/50-PCDT2000 (E).

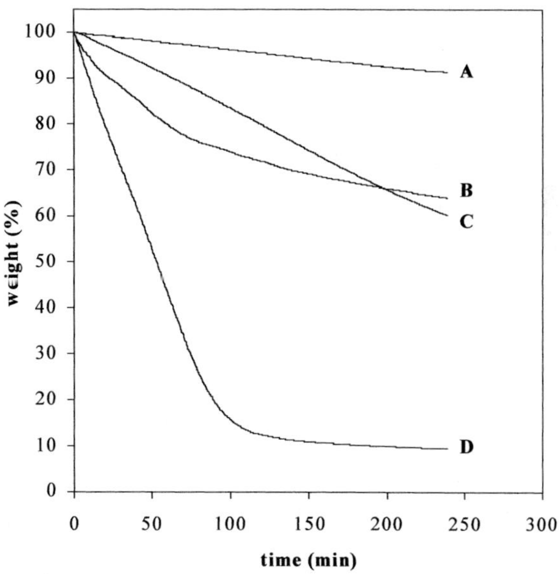

Figure 2. Weight percentage *vs.* heating time at 180°C in air for PLGA50/50 (A), PLGA50/50-PEG2000 (B), PLA-PEG2000 (C) and PEO 200,000 (D).

Viscometric investigations

The molecular weight of copolymers was monitored as a function of the heating time by viscometry. Films of these materials obtained by casting from chloroform were heated in air at 180°C for different periods of time, then they were dissolved in chloroform and their η_{sp}/c was determined. The results are shown in Figure 3 for PLGA-PCDT copolymers and parent homopolymers. Surprisingly, the viscosity of PLGA75/25, PLGA50/50 and PCL exhibits a maximum after around 15 min and then gradually decreases. In contrast, the viscosity of all PLGA-PCDT copolymers strongly decreases in the first 30 minutes and then reaches a plateau. It may be noticed that higher viscosity drops are exhibited by PLGA50/50 copolymers. It was impossible to measure the viscosity of PLGA50/50-PCDT1250 and PLGA50/50-PCDT2000 for t = 240 min since they could not be dissolved. Most probably some degree of crosslinking occurred in the above materials upon heating. A comparison of the FT-IR spectra of crosslinked (t = 240 min) and un-crosslinked (t = 120 min) materials did not allow to gather evidences about the crosslinking mechanism.

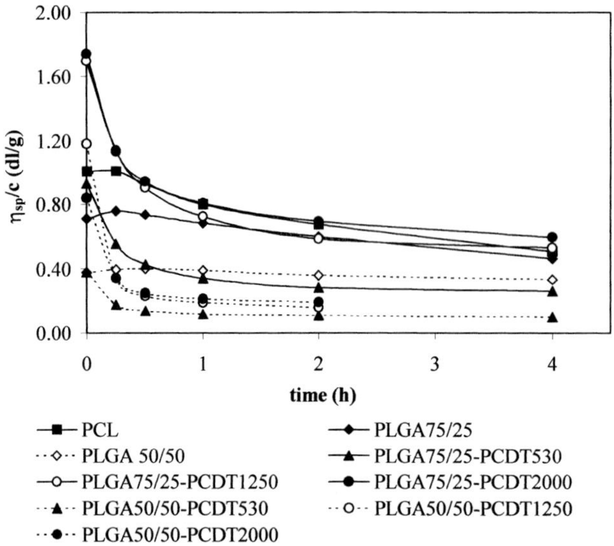

Figure 3. Viscosity *vs.* heating time for PLGA-PCDT copolymers treated at 180°C in air.

The initial increase in viscosity found for commercial homopolymers might be explained supposing that volatile products are already present in these samples and are lost owing to the high temperature. Alternatively, degradative reactions leading to the simultaneous formation of higher molecular weight chains and volatile oligomers might occur, as reported by Watchsen *et al.* for oligomeric PLLA [20]. To assess the presence of volatile products, films of the commercial samples were dried *in vacuo* at 0.1 torr and 50°C for 1 week, then their viscosity was determined. A slight increase in viscosity was observed but it was lower than in heated samples, therefore neither hypothesis could be discarded.

An appreciable WL does not accompany the viscosity decrease in PCL even if elimination of caprolactone has been reported for PCL pyrolised by mass spectrometry [21].

The PLGA75/25-PCDT copolymers are more stable than PLGA50/50-PCDT. Anyway, all multiblock copolymers, when heated at 180°C, undergo a dramatic viscosity drop in spite of the relatively good stability exhibited by the polymers constituting their blocks. This behaviour is probably the consequence of the weakness of chemical bonds linking the segments, which are bound each other alternatively through an ester and a carbonate group. The blocks being themselves polyesters, it seems unlikely that the introduction of a further ester function can affect the stability so dramatically and therefore it seems reasonable to suppose that the carbonate groups constitute "weak points" where degradative phenomena can start.

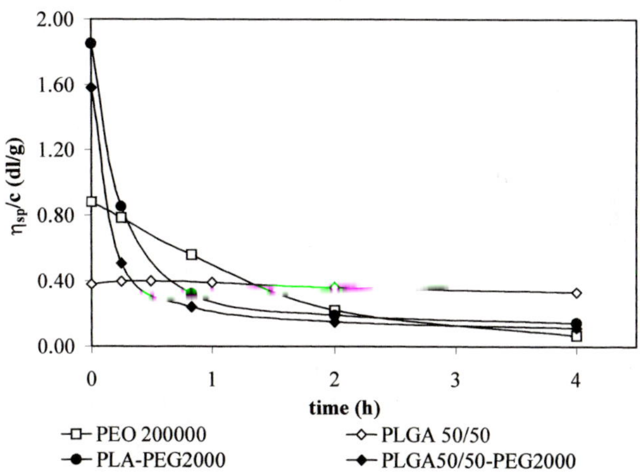

Figure 4. Viscosity *vs.* heating time for PLGA-PEG copolymers treated at 180°C in air.

A similar trend is observed for PLGA-PEG copolymers (Figure 4). The viscosity of PEO decreases considerably and almost linearly over the first two hours, but a much more rapid drop is exhibited by PLA-PEG2000 and PLGA50/50-PEG2000. As in the case of PLGA-PCDT, it is initially more pronounced for the copolymer with the higher GA content.

Nevertheless, while in that case relatively high viscosity values are found for the PLGA75/25-PCDT copolymers even after four hours, in this case the viscosity of both copolymers is drastically reduced within the first hour. This is probably a consequence of the higher susceptibility to thermoxidative degradation of the PEG segments compared to the PCDT segments and it is in agreement with the findings of thermogravimetric analysis.

FT-IR analysis

FT-IR spectra of films treated at 180°C were recorded to monitor possible modifications of polymer structure, which might provide some insight into the mechanism of polymer degradation. PLGA and PCL spectra of samples annealed at 180°C for 240 min are superimposible on those of original polymers. The same happens with PLGA75/25-PCDT but not with PLGA50/50-PCDT copolymers. In the latter, in fact, there are evidences for a preferential degradation of PLGA in comparison with the PCDT segments. For instance, as reported in Figure 5, in PLGA50/50-PCDT2000 the intensity of the PLGA carbonyl band (band A, 1760 cm^{-1}) decreases with the heating time compared to that of PCDT (band B, 1726 cm^{-1}). In PLGA75/25-PCDT2000 no evidence for such preferential degradation of PLGA segments is found, the relative height of the bands being unaffected by heating.

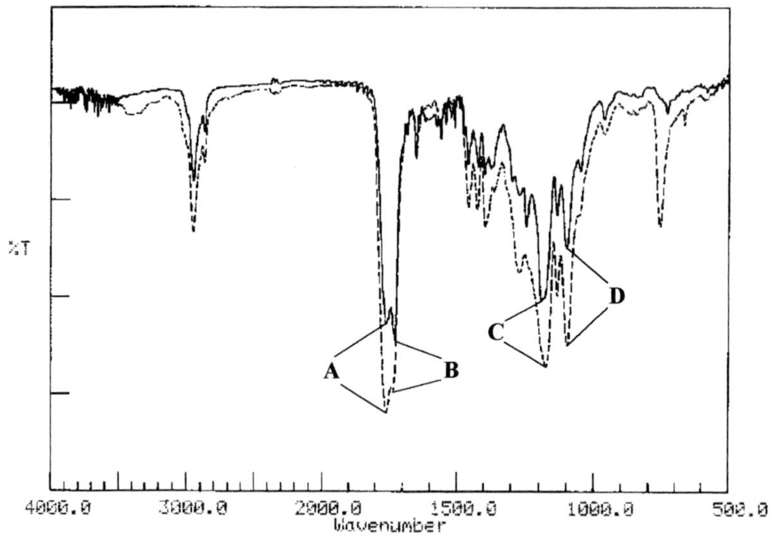

Figure 5. FT-IR spectrum of native PLGA50/50-PCDT2000 (broken line) and of the same polymer after heating in air at 180°C for 4 hours (solid line).

In the copolymers containing shorter PCDT blocks only a single carbonyl band is observed at 1760 cm^{-1}, and therefore it is unsuitable to monitor degradative phenomena. To that purpose, however, the bands located at 1175 cm^{-1} (band C, CO-O stretching of PCDT and PLGA) and 1095 cm^{-1} (band D, O-C-C stretching of PLGA) can be considered. If their relative intensity in samples heated for 240 min is compared with that of the unheated samples, it can be observed that, while in all PLGA75/25-PCDT copolymers no appreciable difference exists, in the PLGA50/50-PCDT samples the intensity of the band at 1095 cm^{-1} decreases relative to that at 1175 cm^{-1} after the thermal treatment. Therefore the FT-IR investigations demonstrate that in the PLGA50/50-PCDT multi-block copolymers the PLGA50/50 segments have a greater susceptibility than the PCDT segments towards thermoxidative degradation.

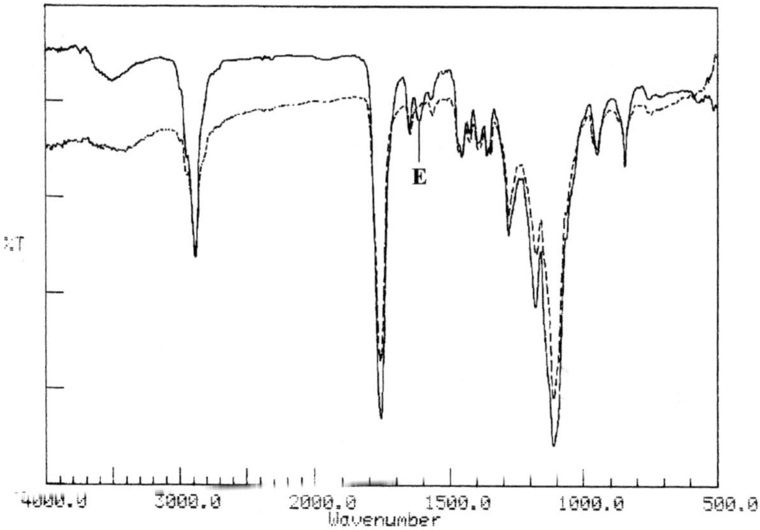

Figure 6. FT-IR spectrum of native PLGA50/50-PEG2000 (broken line) and of the same polymer after heating in air at 180°C for 4 hours (solid line).

A different behaviour can be inferred from the analysis of the spectra of PLGA-PEG copolymers. The most important modification in the spectrum of PLGA50/50-PEG2000 after thermal treatment is the appearance of an absorption at 1613 cm^{-1} (band E in Figure 6). This may suggest that PEG segments degrade forming vinyl ether end-groups, in agreement with the mechanism proposed for the degradation of ethylene glycol based polyesters and copolyesters [19]. No evidence for the presence of such functions can be found in the spectrum of thermally treated PLA-PEG2000, which presents the same bands as untreated PLA-PEG2000. Therefore it is evident that, as in the case of PLGA-PCDT, the LA/GA ratio has a large influence on the degradation mechanism of these copolymers.

Conclusions

The isothermal investigations run in air at 180°C on the PLGA-PCDT and PLGA-PEG copolymers showed that these materials undergo degradative processes which are largely affected both by the type of segments and by the LA/GA ratio. In general, the stability is higher for the PLGA-PCDT copolymers and it increases with the LA content. A rapid molecular weight drop, which may be a consequence of the weakness of carbonate bonds connecting the blocks, takes place in all copolymers upon heating. The FT-IR analysis suggests that, depending on the composition of the PLGA segments, the degradative processes can lead to a preferential degradation of the blocks. Further studies will be necessary to elucidate how the LA/GA ratio can so largely affect the thermodegradative behaviour of these materials.

[1] T. H. Barrows, in: *"High Performance Biomaterials, a Comprehensive Guide to Medical and Pharmaceutical Applications"*, M. Szycher Ed., Technomics Publishing Co., Lancaster, Pennsylvania, Usa 1991, p.243.
[2] S. Li, H. Garreau, M. Vert, *J. Mater. Sci. Mater. Med.* **1991**, 1, 123.
[3] K. A. Athanasiou, C. M., Agrawal, F. A. Barber, S. S. Burkhart, *Arthroscopy* **1998**, 14, 726.
[4] M. Luck, K. F. Pistel, Y. X. Li, T. Blunk, R. H. Muller, T. Kissel, *J. Control. Release* **1998**, 55, 107.
[5] M. S. Widmer, P. K. Gupta, L. C. Lu, R. K. Meszlenyi, G. R. D. Evans, K. Brandt, T. Savelt, A. Gurlek, C. W. Patrick, A. G. Mikos, *Biomaterials* **1998**,19, 1945.
[6] D.W. Grijpma, A.J. Pennings, *Macromol. Chem. Phys.* **1994**, 195, 1633.
[7] M. Ranchandani, M. Pankaskie, D. Robinson, *J. Control. Release* **1997**, 43, 161.
[8] S. Gogolewski, P. Mainil-Varlet, *Biomaterials* **1997**, 18, 251.
[9] D. H. S. Ramkumar, M. Bhattacharya, *Polym. Eng. Sci.* **1998**, 38, 1426.
[10] A. Babanalbandi, D. J. T. Hill, A. K. Whittaker, *Polym. Degrad. Stabil.* **1997**, 58, 203.
[11] M. Agrawal, D. Huang; J. P. Schmitz, K. A. Athanasiou, *Tissue Eng.* **1997**, 3, 345.
[12] M. Penco, E. Ranucci, F. Bignotti, P. Ferruti, *Macromol. Rapid Commun.* **1994**, 15, 683.
[13] P. Ferruti, M. Penco, P. D'Addato, E. Ranucci, R. Deghenghi, *Biomaterials* **1995**, 16, 1.
[14] M. Penco, R. Donetti, R. Mendichi, P. Ferruti, *Macromol. Chem. Phys.* **1998**, 199, 1737.
[15] M. Penco, L. Sartore, F. Bignotti, S. D'Antone, L. Di Landro, *Europ. Polym. J.* **2000**, 36, 901.
[16] M. Penco, F. Bignotti, L. Sartore, S. D'Antone, A. D'Amore, *J. Appl. Polym. Sci.* **2000**, 78, 1721.
[17] S. D'Antone, F. Bignotti, L. Sartore, A. D'Amore, G. Spagnoli, M. Penco, *Polym. Degrad. Stabil.*, in press.
[18] S. Han, C. Kim, D. Kwon, *Polymer* **1997**, 38, 317.
[19] K. C. Khemani, *Polym. Degrad. Stabil.* **2000**, 67, 91.
[20] O. Wachsen, K.H. Reichert, R.P. Krüger, H. Much, G. Schulz, *Polym. Degrad. Stabil.* **1997**, 55, 225.
[21] I. Lüderwald, *Makromol. Chem.* **1977**, 178, 2603.